Carbon Neutral by 2020

Carbon Neutral by 2020
How New Zealanders can tackle climate change

Edited by Niki Harré and Quentin D. Atkinson

craig potton publishing

First published in 2007 by Craig Potton Publishing
98 Vickerman Street, PO Box 555, Nelson, New Zealand
www.craigpotton.co.nz

© Individual authors

ISBN: 978-1-877333-69-9

Printed by Astra Print, Wellington, New Zealand

This book is copyright. Apart from any fair dealing for the purposes of private study, research, criticism or review, as permitted by the Copyright Act, no part may be reproduced by any process without the permission of the publishers

CONTENTS

Introduction
  A call to action — 6
  New Zealand's carbon crisis *Quentin D. Atkinson* — 9
  The psychological challenge of climate change *Niki Harré* — 15

Schools in a carbon neutral world — 28
  Heidi Mardon

A new paradigm for home renovation — 39
  Jette de Jong and Niel de Jong

Carbon neutral living in the typical New Zealand house — 78
  Brenda and Robert Vale

Thinking outside the car: how we can achieve carbon neutral transport — 97
  Julie Anne Genter

Reducing the carbon burden from Auckland's transport system — 113
  Ning Huang, Penelope Anson and Robert Vale

How 'Hobson Mall' became climate and people friendly — 131
  Maggie Lawton and Robert Vale

Computing away climate change — 144
  Alexei Drummond, John Hosking, Christof Lutteroth, Gerald Weber and Burkhard Wünsche

Deep organics — 164
  Brendan J. Hoare and Keith Thomas

Back to the drawing board: sustainable design — 185
  David Trubridge

The role of ethics in climate change strategies — 199
  Prue Taylor

Responsible investing — 216
  Rodger Spiller

The sustainable business challenge — 235
  Rachel Brown

Carbon neutrality and the law — 258
  Klaus Bosselmann

Political activism and carbon neutrality — 276
  Jennifer Curtin and Anita Lacey

Conclusion — 292

Appendices — 295

Index — 302

Introduction

A call to action

The crisis is clear. The globe is warming and our environment is changing. Ice caps are melting, sea levels are rising, species are disappearing and the weather is getting wilder. Soon, droves of people may be on the move to escape lands made barren by the effects of climate change.

Paul Slovic, a psychologist who has studied risk perception, discovered that people are most afraid of things that are dreaded and unknown.[1] Because of this, it takes much less to trigger gut-level reactions to the threat of a dirty bomb or a nuclear power station meltdown than it does to the prospect of a few extra degrees in temperature. After all, the slowly accumulating effects of global warming seem mostly to be just more of what we are used to – more heat waves, more water shortages, more deserts, more coastal erosion. Environmental organisations, climate scientists and journalists have worked hard to get across the message that the current trends in temperature are out of the ordinary. We have seen images of polar bears stranded on floating ice, acres of baked earth riddled with cracks, rubble in place of glaciers and children clinging to coconut palms as waves crash over their village. Of late, even the economists have entered the debate, with the Stern Review[2] predicting billions of dollars in lost earnings and costly repairs if we do nothing.

Despite the odd distraction such as sunspots and previous hot periods, the great majority of experts are telling us that climate change is primarily the result of greenhouse gases produced by Western lifestyles. Releasing the carbon from oil, gas and coal trapped underground for millions of years in a mere century or two and cutting down forests which previously provided sinks for carbon are major contributors. Some, but not all, experts say we may be able to stop the worst effects if we change these practices now.

For many of us, the reality has sunk in – climate change is happening, it has almost certainly been caused by us, and if we do not do something about it now it is likely to change the face of the planet. We are at least a little scared, especially if we start to imagine the world our children and grandchildren will inherit, but we are also confused.

At the core of our confusion is one nagging question: Whose problem is it?

It is tempting to think it is the government's problem. In New Zealand, we may even feel that it is the problem of other governments. Unlike many problems, we cannot solve it for ourselves, because it is about the atmosphere we share with everyone else. No matter how good my family, my organisation, my city or New Zealand is at reducing greenhouse gas emissions, bad things will still happen to us if people all over the planet don't act similarly. So let's leave it to the USA or the European Union – they are the real offenders and the ones with the power to solve the problem. If New Zealand can and should do anything, it is really up to the politicians in Wellington to make the necessary changes. Climate change is big – and big things can only be solved at the highest level.

There are two downsides to assuming it is all the government's problem. One downside is that all the governments we are talking about are elected. In democratic societies, politicians must tread a fine line between doing what they believe is best for the country and doing what they believe will get them re-elected. There is little point in becoming the government, bringing in unpopular laws, losing the next election and then watching the other side repeal all your good work. Not to mention the fact that deeply unpopular laws would probably be unworkable as people are extraordinarily creative at sabotaging rules they don't like.

So, if everyone continues to take full advantage of the plastic bags, plasma TVs, cheap beef, takeaway coffee cups, imported oranges, microwave meals, big vehicles, air travel and apparently limitless landfill on offer, politicians will rightly figure that people are just not ready to change the way they live. If no one writes letters to newspapers saying they are worried about the state of the planet, if no one signs climate change petitions, if no one takes the bus or rides their bike, if no one buys cloth nappies or carbon credits, if no one runs a carbon-neutral business, if no one plants native trees, politicians will get the message that we like things the way they are. And, given that message, they will be reluctant to do much.

A second weakness with leaving it to the politicians is that you'll miss out on the opportunity to be part of one of the greatest issues of our time. In 2006, Helen Clark announced the government's aim for New Zealand to be carbon neutral. Whatever your views on the politics behind this announcement, that our prime minister has publicly declared this as a national aim is extremely important. Imagine living in a country united by an effort to protect the planet for future generations. It is hard to envisage a more positive basis for our national identity.

This book has been written by a collection of people keen to be part of the solution. They are all experts from different fields, who have been given the task of devising strategies for New Zealanders to work together to become carbon neutral by 2020. The result is 14 chapters rich with ideas they hope will inspire you to get started – or keep going – with the collective effort to prevent the worst-case scenarios for climate change from becoming a reality. The authors do not all agree on what is best, but they do all agree that it is absolutely worth trying to create a better future. For most of the contributors to this book, climate change is just one of many pressing issues involved in trying to live sustainably on this planet. They are also concerned with human wellbeing and the health of the planet in a much wider sense. Climate change, to many of us, is not only a critical challenge in itself but a wake-up call that invites us to look more broadly at all the problems our current lifestyles are creating.

So, the book does more than just present solutions to the problem of climate change. It also presents visions for a better society, one that is cleaner and fairer and encourages people to be their best. The authors have tried to be realistic in their visions – they understand that you cannot unpick all the deeply interwoven practices and ways of thinking that have got us into this predicament in one fell swoop. But they also understand that you have to take a few risks and have faith that people are willing to give up some of their dubious luxuries for the sake of a healthier world.

First, an overview of New Zealand's carbon emissions is presented. This is followed by a discussion of climate change as a psychological challenge. Subsequent chapters cover schools, housing, transport, malls, computing, organics, design, ethics, investment, business, law and political action. They do not cover every aspect of our lives, but the areas chosen are important for most New Zealanders. Each chapter begins with a short overview of how the authors see New Zealand's current situation, their vision for New Zealand in 2020 and the strategies they propose to allow us to achieve that vision. The book finishes with a conclusion that urges all of us – as individuals, members of organisations and political agents – to take action now.

New Zealand's carbon crisis
Quentin D. Atkinson

Quentin D. Atkinson has a PhD in psychology from the University of Auckland. He is currently a research fellow in the Institute of Social and Cultural Anthropology at the University of Oxford, where he uses mathematical modelling techniques to investigate the evolutionary history of human genes and culture. His research interests include the evolution of religion, the feasibility of large-scale group cooperation, and the evolution of languages in Central America, the Pacific, Europe and the Near East. He has a long-standing interest in environmental ethics and sustainability, and in 2005 represented New Zealand at the World Youth Forum on Sustainable Development.

Greenhouses are great for keeping things warm. The Earth's atmosphere is like an enormous greenhouse around us, keeping temperatures on the ground habitable. This is called the 'greenhouse effect' and it arises because certain gases in the atmosphere, such as carbon dioxide (CO_2), methane and water vapour, act like the glass walls of a greenhouse, trapping the sun's energy and keeping the planet warmer than it would otherwise be. Global climate models show, not surprisingly, that the strength of this effect depends on the amount of greenhouse gases in the atmosphere. We can see evidence of this in microscopic pockets of air trapped in ice core samples from the polar ice sheets. These layers of ancient ice, going back hundreds of thousands of years, reveal that as the concentration of CO_2 and methane rises and falls, so too does the Earth's temperature.[3]

The reason greenhouse gases are now a problem is that over the last

150 years, and particularly the last 50 years, their concentration in the atmosphere has risen dramatically. For millions of years the amount of CO_2 in the atmosphere has been held relatively constant by the carbon cycle – CO_2 is released by animals, plants and microbes when they respire, and absorbed by plants during photosynthesis. However, since the industrial revolution, humans have been burning vast quantities of fossil fuels and cutting down huge areas of forest. Fossil fuels, such as the coal we use to run our power stations, the gas we use to heat our homes and cook with, and the oil we run our cars on, all release CO_2 and other greenhouse gases when they are burned. And the forests, which would normally be able to absorb large amounts of CO_2, are rapidly disappearing.

The best computer models of the global climate indicate that by changing the mix of gases in the atmosphere, we are enhancing the greenhouse effect and causing the Earth to warm significantly. Since the industrial revolution, atmospheric CO_2 levels have increased by over 35 per cent, and global average annual temperatures have risen by roughly 0.6°C.[4] This may not sound like much, but already glaciers all over the world are retreating,[5] the sea level is rising,[6] severe weather is more frequent,[7] coral reefs are bleaching,[8] and the sea ice in the Arctic and Antarctic is shrinking annually,[9] threatening species like polar bears and walrus with extinction. More worrying though, is the prediction by the Intergovernmental Panel on Climate Change (IPCC) in 2001 that, if we do not change what we are doing, by the end of the century, CO_2 levels will more than double and global temperatures will rise by between 1.4°C and 5.8°C, with potentially dramatic effects on the Earth's ecosystems.

If this sounds like a long way off and not such a big change in temperature, consider three things. First, 90 or so years is not a very long time in the greater scheme of things – Sir Edmund Hillary was born just under 90 years ago! Even if we will not be around then, many of us have children or grandchildren who will be, and the effects of climate change are not going to be stored up until the end of the century – they are already being felt. If the 5.8°C temperature increase proves correct, then we can expect the kind of changes that we have seen over the last 150 years (with a 0.6°C rise) to occur every decade, probably with increasingly serious consequences. Second, the average temperature during the last ice age was only 5-9°C cooler than today's average,[3] but the effect was huge. Global sea levels, for example, were so low that it would have been possible to walk from Australia to Papua New Guinea and from Russia to Alaska. And finally, the 1.4°C and 5.8°C estimates for warming do not take into account what are called 'positive feedback loops', in which the effects of warming cause more warming. For example, warming beyond 2°C may cause the Amazon rainforest to dry out and die.[10] The vegetation of the Amazon absorbs vast quantities of CO_2 and, without it, atmospheric CO_2 levels will rise even faster, causing more warming more quickly. By accounting for these kinds of positive feedback loops, some scientists predict temperature rises of up to 11°C by the end of the century![11]

In June 2005, the world's leading scientists published a statement asserting that the evidence for climate change is clear enough that we must take prompt action to reduce and mitigate emissions of greenhouse gases.[12] As I write this chapter, two years on, the evidence is even clearer, but there is a worrying lack of action in New Zealand and abroad.

Faced with the world's biggest environmental crisis, as a New Zealander it is easy to feel complacent or even smug. New Zealand enjoys a clean and green image. We like to think of ourselves as eco-friendly and we sell this label to the rest of the world. Indeed, our contribution to global emissions is small – New Zealand produces less than half a per cent of global greenhouse gas emissions.[13] So, it is tempting to point the finger at the big polluters – most notably, the USA, Europe and China. However, our small impact is really just a product of our small population. Actually, on a per person basis, there are only 10 nations in the world that emit more greenhouse gases than we do. Like our favourite sport, when it comes to pumping out carbon, we punch above our weight.

New Zealanders emit annually an average 19.4 metric tonnes of CO_2 equivalent per person.[14] If that sounds like a lot, that is because it is. The global average is roughly four tonnes of CO_2 equivalent per person, but even this is unacceptably high. In his book *Heat*,[15] the British journalist George Monbiot proposes that, at the very least, we should be trying to prevent global average temperatures from increasing by more than 2°C, beyond which, positive feedback loops, such as destruction of the Amazon rainforest and the release of methane from melting Siberian permafrost, could cause global temperatures to spiral even higher. To prevent a 2°C increase in temperature, climate models predict that our fair share of CO_2 emissions (dividing the acceptable total emissions by the number of people on the planet) is less than one metric tonne per person. This means that we New Zealanders are each producing over 20 times more greenhouse gases annually than is sustainable.

A major reason why New Zealand emits such large quantities of greenhouse gases is what we choose to do with our land. Prior to the arrival of humans (circa 1000 AD), forests covered 80 per cent of New Zealand – effectively everywhere that trees could grow. Uncontrolled fires lit by the early Maori, and later deforestation by European settlers has left us with only 24 per cent of our land area forested. The forests we have left can absorb only about 33 per cent of our national greenhouse gas emissions. If that was not bad enough, most of the land that was once forested is now used to farm ruminants like sheep and cattle. As they digest their food, ruminants belch out methane, a potent greenhouse gas, 25 times more powerful than CO_2. As a result, agriculture accounts for roughly half of New Zealand's greenhouse gas emissions.

Clearly we are going to have to think hard about how we use our land and the viability of continuing to farm animals in this way at such a high environmental cost. However, in this book we have chosen to focus on the other half of New Zealand's greenhouse gas problem – urban city dwellers. Eighty-six per cent of New Zealanders live in an urban setting. Even ignoring the half of our greenhouse gas emissions due to methane from farm animals, our urban lifestyles mean that New Zealanders still produce more CO_2 per person than, for example, the British, and 10 times more than our fair share globally.

So how do we manage to emit so much CO_2? New Zealand is lucky in that, unlike most countries, two-thirds of our electricity is generated by renewable, carbon-neutral resources – mostly thanks to our hydro dams. However, the remaining one-third of our electricity generating capacity comes from non-renewable coal and gas-fired power stations, which together account for about one-sixth of New Zealand's CO_2 emissions. So, despite our renewable energy resources, powering our homes, schools, offices and factories accounts for a substantial portion of our greenhouse gas emissions.

Kiwi homes are particularly energy inefficient. Over one-third of New Zealand's electricity is used in our homes to power our lights, heating, fridges, freezers, air conditioners, TVs, DVD players, computers, washing machines, playstations and whatever other gadgets we have managed to accumulate. Our home lives also produce CO_2 emissions in less direct ways. Like citizens of other Western, industrialised countries, New Zealanders like to buy 'stuff', and, as a general rule, the more we buy, the greater our CO_2 emissions. This is because we burn or refine fossil fuels to produce most of the energy and many of the raw materials required to manufacture and distribute goods and services. Today, almost every product or service we use causes CO_2 emissions during its manufacture and distribution. This now applies even to the food we grow. Farm machinery and fertilisers all produce CO_2 in their manufacture and use, as does the packaging, refrigeration and lighting in the supermarkets that sell our food. Even more CO_2 is produced if products have to be shipped or, even worse, flown long distances before they are sold.

Another major contributor to our CO_2 emissions comes from carting people around. New Zealanders love their cars. In fact, New Zealand now has more cars per person than almost any other country in the world.[16] There are 62 cars for every 100 people living in New Zealand.[16] This is even more astounding when you consider that a substantial proportion of our population are either too young or too old to drive. In fact, we probably have about one car for every person who is eligible to drive. Worse still, cars have become the preferred mode of transport, even for very short trips – one-third of car trips are less than two kilometres and two-thirds are less than five kilometres,[16] so most of the time we could be walking or cycling. All of this means that land transport accounts for almost 36 per cent of New Zealand's CO_2 emissions.

We are also travelling by air more than ever before. Currently air travel accounts for just 2.6 per cent of our CO_2 emissions, but this figure is rising rapidly. Globally the contribution of air travel to emissions is small, but only because it is an activity that a mere fraction of the world's population indulges in. One return flight from Auckland to a European city produces about 10 metric tonnes of CO_2 per passenger[17]– that is about half the average New Zealander's annual emissions and 10 times our fair share. It is even worse than this though, because the contribution of greenhouse gas emissions due to air travel is exacerbated by the fact that CO_2 emitted at high altitude has three times the effect on global warming.[18]

We are left with a choice – to make the changes that climate scientists tell us are essential, or to stick our heads in the sand and hope that the rising waters will not drown us. It is true that there is some uncertainty about exactly what will happen if we fail to take action. It is also true that, owing to the time lag in the climate system, the greenhouse gases we have released into the atmosphere over the last decades have already committed us to a certain amount of warming. But if New Zealand and other countries do nothing to reduce emissions, we are essentially playing Russian roulette with the climate, inviting a host of potentially disastrous environmental changes. Different parts of the world are likely to be affected differently. Those from Third World countries, who have had the least to do with creating the climate crisis, are likely to be the worst affected and the least well equipped to deal with the consequences. But we can be sure that New Zealand will also be affected. The Ministry for the Environment[13] predicts that New Zealand will experience rising sea levels, increasing storm surges, erosion and flooding in coastal lowland areas, including many of our coastal towns and cities. The prevailing westerly winds will become stronger across the country and cyclones will become more frequent, especially in the North Island. Drought-prone areas, especially along the east coast of the North and South islands, are likely to get drier and flood-prone areas will experience more extreme flooding more often. None of this considers the effects of the more severe temperature increases that have been proposed.

Recognising the seriousness of the situation for New Zealand and the world, the New Zealand government has been relatively progressive in its climate change policy, although less progressive in its implementation. In 2002, we became one of 165 nations to sign the Kyoto Protocol – a legally binding international agreement obliging us to reduce our CO_2 emissions to 1990 levels by 2012. Although this commitment is nothing like the 95 per cent emissions reduction we require, many view it as an important first step towards an international strategy to tackle climate change. In 2003, in an effort to allow New Zealand to meet its obligations under the Kyoto Protocol, the government attempted to introduce a levy on emissions from farm animals. The 'fart tax', as it became known, received intense opposition from farmers, and was abandoned in 2003. More recently, New Zealand was to become the first country in the world to use a carbon tax to incorporate the environmental cost of goods and services into their purchase price and hence reduce emissions. This tax was also abandoned in late 2005. The reason the government gave for abandoning the carbon tax was that it was expected to have only a limited effect on reducing emissions. However, a more honest explanation may be that public opinion would not allow the level of taxation that is required to change behaviour. The failure to introduce any effective measures to curb carbon emissions means that

New Zealand is not on target to meet its obligations under the Kyoto Protocol. To offset our excess emissions we will have to buy carbon credits (effectively buying the right to pollute from those countries who do manage to meet their emissions targets), at an estimated cost of roughly NZ$500 million. More importantly, though, we will have failed to do our bit to stop climate change.

As the two failed carbon tax proposals show, the government cannot avert the climate crisis for us. Government policy must reflect public opinion. The government's current climate change strategy, which essentially involves raising awareness of the issues and leading by example, may thus be the most we can expect from our political leaders at the moment. This is because climate change is not just 'the government's problem'; it is our problem. Each of us has to educate ourselves and others about our impact on the climate and begin to work towards changing the way we do things. Legislation will have a role, but not until we demand it. To avoid climate chaos, we are going to have to completely rethink the way we live, work, play and get around.

The psychological challenge of climate change
Niki Harré

Niki Harré is a senior lecturer at the University of Auckland where she has taught social and community psychology for 10 years. She has published many studies on driving and other traffic behaviours. More recently she has been researching psychological wellbeing, positive youth development and political activism. Niki is a founding member of SALT, a community group in her suburb of Pt Chevalier in Auckland, aimed at encouraging people to walk and cycle.

People's brains have to work hard these days. The modern world barrages us with problems that tax our cognitive resources to the limit. Climate change is one such problem. To really accept that the globe is warming owing to human activities and that it is our collective responsibility to take action, we have to get our head around a series of mental obstacles put in place by our evolutionary history and cultural practices. In this section, I will look at these mental obstacles and try and convince you that despite them, there are alternatives to a society in which most people go about business as usual, leaving others to worry about where it is all heading.

Climate change is a theory about how the world works. It also has a moral aspect, as tagged to almost any discussion of it are statements implying that its effects will be 'bad' and that we 'should' do something about it – now. That is fine from a human perspective, as we are adapted to think in theories, and are moral creatures intensely concerned with right and wrong. However, this brings us to the first psychological obstacle to coming to grips with climate change: it bears almost no direct relationship to our personal experience, especially in New Zealand. Personal experience is the easiest way to learn and information absorbed in this way tends to influence our behaviour more readily than information that we are simply told is true.[19] For example, we eventually catch on that it is colder in winter than in summer by living through the seasons and feeling the difference for ourselves. Once we are grown-up and sensible, we do not need to be persuaded to wear shoes outside in July – distant memories of that cold footpath are enough to make us do so without prompting.

On the other hand, there is nothing for most of us to actually see or feel about global warming. We must have faith in the scientists who take measures from ice cores and tree rings and put the information into computer models, producing graphs that we are told add up to big trouble. We must stay staunch in our faith when another person who also professes to be an expert comes along and tells us these graphs show a cyclical blip which adds up to nothing much. If we decide to offset our carbon emissions by putting money towards a tree-planting scheme, we must simply hope that this is the right thing to do – in the face of ongoing debates which most of us do not understand about the value of carbon offsetting. For most people, the reality (or not) of climate change, and its causes and cures, are conversations being had by experts, in which we must take sides. Not only does this make it easy to bounce from one view to another but it also enables us to disengage in a way that we cannot so easily do with issues we have learnt about first hand (like footpaths in winter).

Environmentalists know that people respond more readily when they are witnesses to an event, and so have produced the next best thing – images of what other people have witnessed. Footage of glaciers cracking and plunging into the sea fills up the visual and auditory centres in the brain and can almost make us feel that we have seen climate change at work. Photos of people suffering in hurricanes or heat waves not only take us into the event, but also engage one of our more endearing emotions – empathy. In his autobiography, Bob Geldof [20] attributes the readiness of rock stars to participate in the charity group Band Aid as being due to television coverage of people starving in famine stricken Ethiopia. For Geldof and others, it was these images that shifted them from an abstract understanding that famine was occurring to a realisation of exactly what famine meant in human terms. So, although the idea of the globe warming might not be intrinsically alarming to the human mind, pictures of climatic catastrophes and the associated human stories may do a lot to shake us out of our complacency. It strikes me that one of the more useful things a filmmaker could do in New Zealand would be to show exactly what will happen here by 2100 if warming trends continue unabated and sea levels rise as predicted. I think for many people, images of an eroded cliff where their favourite beach used to be may bring home the message that climate change is happening here too.

Another tricky thing about climate change is that its impact is global. It has long been established that the more people there are to take action to rectify a problem, the less likely it is that any particular individual will do so. A series of experiments done in the 1960s and '70s demonstrated this phenomenon, labelled the 'bystander effect', with groups of between three and six individuals.[21] With climate change, there are potentially 6 billion others out there who could do something. Can you get away with leaving it up to others? Absolutely. That way you will avoid looking like an idiot for caring too much and, even worse, doing more than your share. Unfortunately, this kind of thinking, which comes rather easily to most of us, turns into a vicious cycle, as we all ignore the problem for now, looking to others – who are also ignoring the problem – to see what we should do. Interestingly, despite the powerful human tendency to conform, sometimes even in the face of glaring evidence that the behaviour we are conforming to is clearly wrong, it may take only one person to break the cycle and open the door to alternative responses.[22]

Despite popular rhetoric that human beings are consumed by self-interest, people do care about others. There is abundant evidence for altruism, that is, acts that benefit someone else at a cost to the altruistic individual (time or energy, for example). Even in little ways, we see this all the time, when people bake a cake for the neighbours who have just had a baby, give money to the City Mission, help a visitor find the office of a co-worker. However, altruism is more evident among people that are close to each other. We are most likely to put ourselves out for our children, other relatives and people that are clearly similar to ourselves.[23]

This leads to a second way in which climate change's global nature is an obstacle to action. We are equipped to care, but not for everyone. In fact, we have probably evolved the ability to readily distinguish between 'our' people and 'other' people. Those who are part of our group, be it our family, our school, our workplace, our country, are included in our map of ourselves. They are a critical part of our identity, signifying where we fit in the social world.[24] This group-based thinking is part of the reason why in certain historical circumstances ordinary people have needed little persuasion to fight wars, keep slaves and ignore large-scale human misery. In each case, the people we kill, own or allow to suffer and die are not included in those groups that make up the social aspect of our identity. They are others, and the needs of others can be overlooked when it comes to our own survival and yearning for social status, comfort and novelty. Sometimes we are driven by hate, but it is mostly because our minds are so full of the plans we have for ourselves, and those we care for, that any concern for the suffering of others just fades away. The rock stars who contributed to Band Aid were temporarily jarred out of this state by having the suffering of others thrust at them in a new and shocking way. Eventually, however, they mostly slid back into lifestyles that put themselves and those closest to them first (Geldof himself excepted).

Climate change is about everyone, which makes accepting it a pretty tough call for the human brain. Could we take people's group-based thinking and modify it, by persuading individuals to identify with something as large as humanity? One passage that makes me feel part of a glorious all-in-it-together, human community is from an essay by Tony Kushner.[25] At the beginning of the essay, the writer is told by a taxi driver that there is a supernova 60 light years away, destined to wipe out the Earth. The essay then goes on to describe the more likely ways in which we will be 'got', including global warming, the collapse of our economic system or a catastrophic war, and urges people to play a part in solving these problems. At the end of the essay, he puts the following challenge:

So when the supernova comes to get us we don't want to be disappointed in ourselves. We should hope to be able to say proudly to the supernova, that angel of death, 'Hello supernova, we have been expecting you, we know all about you, because in our schools we teach science not creationism, and so we have been expecting you, everywhere everyone has been expecting you, except Texas. And we would like to say, supernova, in the moment before we are returned by your protean fire to our previous inchoate state, clouds of incandescent atomic vapour, we'd like to declare that we have tried our best and worked hard to make a good and just and free and peaceful world, a world that is better for our having been here, at least we believe it is.'

The movie *An Inconvenient Truth*[26] used many images that drew attention to the planet everyone shares, sometimes portraying greenhouse gases, rather like the supernova in Kushner's essay, as a common enemy. This is different from the usual rhetoric (employed at the beginning of this book), that climate change is our fault. Even if it is our fault, talking about climate change as something external that we can and should join together to overcome can be very empowering. Guilty burdens weigh us down, but fearful enemies inspire action.

The Make Poverty History campaign has used the Internet to cut across geographically-based group affiliations and encourage people to care about social justice on a global scale. Interestingly, however, the websites themselves are still nationally based and primarily designed to appeal to citizens of a particular country. So, the New Zealand site[27] starts by saying 'Thousands of New Zealanders have joined with millions around the world calling for an end to poverty.' And at the time of writing, the USA-based 'One Campaign'[28] is focused on its country's 2008 presidential elections. The national focus of these websites may be mostly a practical necessity, because the world is divided into groups in a real sense, but it is also because people have ready-made national identities that can be instantly tapped into. Working with group-level identities in overcoming climate change will be taken up again later.

Not only is climate change global, but it is also long term. Early studies on pigeons and children showed that both are willing to reduce the size of the reward they will receive in order to get it immediately.[29] Having somewhat better self-control, adults are a little more patient, but our patience has limits. Going on a diet for two months in order to fit into summer clothes? Maybe. Putting aside money each week for retirement? Only if someone puts a form in front of us and then takes the money before we can get our hands on it. Doing actions now for the sake of future generations? A challenge indeed. But this response is not as selfish as it seems. The longer the time frame, the higher the risk that our well-intentioned actions will fail to achieve the intended aim, and people sense this. Most of us would like to leave the world a better place than we found it; the problem is that it is just so hard to know if what we do now will meaningfully contribute to this end. Like jumpy Wall Street investors, uncertainty puts us off.

I have left the biggest, most daunting mental obstacle for last. This is the one that almost certainly is affecting you right now, even if you are thoroughly convinced that the globe is warming, feel it is your responsibility to do something about it and care deeply about present and future people all over the planet. It is the huge social conspiracy that we all create and yet may be only dimly aware of – the conspiracy to keep things just as they are. Take my mornings. I wake up to a battery-run alarm clock, turn on an electric light, open a plastic packet of rolled oats, get a plastic bottle of milk from an electric fridge, make a coffee using a fancy electric machine (sometimes organic, but rarely fair-trade, as I haven't seen it in 'fine-ground'), put on old Asics running shoes (designed in the USA, made in China) and make my lunch (tuna from Thailand out of a tin, lettuce grown in the backyard, paper wrap for the sandwiches, a muesli bar covered with shiny paper made from goodness knows what, New Zealand mandarins). Sometimes, at about the moment I open the rolled oats packet, I feel a momentary sense of helplessness. Here I am, re-creating our environmentally and socially disastrous system by the very first things I do each day.

In theory, the individual could do a lot better; I could do a lot better. In practice, the mental effort to do things that are not encouraged by the social system is too much for most of us most of the time.

First, we are kept hooked into the status quo by habit – we have grown up with things pretty much the way they are now, and learnt ways of managing ourselves that seem to work OK. Circuits in our brain have developed that almost automatically guide us through these much-practiced behaviours.

Second, we are kept hooked in by social obligations and competing values. So, what do you do when your daughter is in a ballet recital?

Like any good parent, you buy her a glittery dress that she will only wear once, drive her to the theatre (three times), give her $5 for another mother to take her to McDonalds after one of the shows and then find it hard to enjoy the performance in quite the way you used to. You do not withdraw her from ballet classes on environmental grounds and not just to avoid the embarrassment. It is because you are part of the social network that goes along with your daughter attending ballet classes. You have easy, friendly interactions with the other parents, and appreciate the time the teachers spend and their enthusiasm. You also truly value what ballet offers – a cultural tradition, musical and physical training, self-discipline.

I have outlined above some key obstacles to people taking climate change on board (or in brain) and hinted at ways to overcome some of them. But more is needed to keep people motivated and encourage collective effort. More is needed, too, if we are going to see climate change as an opportunity to create a new way of thinking about the planet and our place on it, rather than just as a technical problem that scientists and governments will eventually solve, we hope. What we need is for the values that underpin a positive response to climate change, and that are already part of most people's psychology, to come to the surface and infuse daily actions. What we also need is for organisations – families, schools, workplaces, sports clubs, ballet schools, cities and countries – to draw up new rules for themselves that put environmental protection to the forefront. What we need, I believe, is for masses of people to have 'climate change identity projects'.

Identity projects are mental blueprints that we carry in our heads and that direct our activities.[30] We all have them – my current ones include being a good mother, editing a carbon-neutral book, running a half-marathon in under 1 hour 50 minutes. Identity projects have a number of characteristics. First, they are directed towards a particular goal. Sometimes the goal is very broad, such as being a good mother, and sometimes it is highly specific, as with my half-marathon time. Second, they involve action – implicit in the term is the idea of projecting ourselves into the world. This distinguishes projects from fantasies, ideas of what we might do one day, if we ever get around to it. Third, they have at least some level of coherence. That is they involve a variety of thoughts, emotions and actions that all stem from efforts to achieve the outcome implicit in the project.

Identity projects come and go. Successful projects tend to grow, becoming more firmly entrenched in our sense of who we are, encroaching on a wider range of activities and sprouting new projects.

Many of us already have an identity project directed towards reducing our environmental footprint and in particular our carbon emissions. I, perhaps like you, consider my carbon footprint in most things I do. I turn out lights, take my bike or walk to most local destinations, agonise over plane trips, contribute to the carboNZero carbon offsetting scheme, rarely accept a plastic bag when shopping and mostly resist new gadgets. However, I also do things that I know others consider unacceptable, like eat tuna, and almost certainly overlook practices, through ignorance, that will one day prove hazardous to the planet. But reducing my carbon output is an identity project I cherish. It makes me feel good about myself and my place in the world, and provides me with direction. My momentary crises, such as the rolled oats-crisis described earlier, are usually about the compromises I have to make to still be part of the social world and to participate in relationships and activities I value. They are rarely about all the wonderful products I am missing out on through my self-imposed rules. In terms of motivation, I am beyond considering whether what I do 'makes a difference'. Instead, I do what I do because of how I want to live and who I want to be.

If this sounds like a testimonial for the benefits of a carbon-conscious identity project in helping you find a fulfilling way through life, it is. Numerous studies in psychology have shown that people need meaning.[31] Meaning may even be our primary need, the one thing we can hold on to when the going gets tough. Meaning can be found through the love we give families and friends, doing a good job at work or spiritual practise. Doing things for others is a particularly powerful source of meaning, and according to the religious scholar Karen Armstrong, is advocated by all major faiths 'because it has been found to be the safest and surest means of attaining enlightenment'. As she goes on to say:

> Self, after all, is our basic problem. When I wake up at three in the morning and ask myself, Why does this have to happen to me? Why cannot I have what X has? Why am I so unloved and unappreciated? – I learn that ego is at the heart of all pain. When I get beyond this for a few moments, I feel enlarged and enhanced.[32]

In interviews that my students[33] and I have conducted with people who have dedicated their lives to the cause of social justice, there is no hint of regret for the opportunities to advance themselves that they may have missed out on. Instead, they talk of the sense of meaning they have achieved in living consistently according to their values.

If you are doing what you do because it feels right for you, you are less vulnerable to the tales of doom-and-gloom in the media and the cynical chuckles of others about your wasted efforts. Paradoxically, however, by riding alone into the prevailing wind (literally, perhaps), you will almost certainly influence others. People are great copiers, and some will copy you. Even by turning out the office light as you leave your room for lunch, you become the person who breaks the leaving-on-the-light conformity cycle. The image of you, flicking off that switch, will – sooner or later – be noticed by someone who is in the right mental space to make that behaviour their own. Eventually, maybe, the practice will spread until leaving the light on when you go out will seem as strange as dropping an empty chip bag in the corridor.

In the space of a few years, climate change has gone from being almost unheard of in mainstream society, to being a possibility, to being a 'fact'. What this means is that it is now at least possible to put climate change and other environmentally related issues on the agenda in most contexts. If having your own carbon-conscious project can be good for you as well as the planet, collective carbon-conscious projects can be even better. Almost all of us are in positions where we could initiate shared carbon-conscious identity projects in one or more of the organisations to which we belong. Recently, I was able to suggest that our local secondary school include working towards environmental sustainability in all areas of school life as a goal for the next three years and not one person spoke in opposition. I have heard other stories, too, of people tentatively suggesting that their organisation take steps towards reducing their greenhouse gas emissions and meeting with unexpected success. I sometimes wonder whether those of us who fancy ourselves as being 'hard core' on these issues do not realise just how mainstream our concerns have become.

So what would a successful, shared identity project around carbon-consciousness look like? To start with, it would ideally be located within one or more of the groups to which people already belong, and so take advantage of the group-based thinking that shapes people's sense of themselves.

The smallest unit up from the individual is usually the household. People who live together tend to grow more similar in their values (or fall out). There is a lot to be said for hurrying this process along somewhat through making explicit the household values. In his book *The Seven Habits of Highly Effective Families*, Steven Covey[34] advocates family meetings and weekly planning as keys to harmonious family life. Family (or household) meetings can be ideal places to shape a shared identity project around something that is meaningful to you – be it climate change, environmental sustainability or just doing your bit to create a better world. One possibility is to choose an action each month that the household could do. Children are often easily moved by animal welfare issues – buying only free-range eggs was a shift we made a few years ago because of a family meeting. At this level, and particularly with children, it does not matter if the project is abandoned from time to time nor if the actions that you collectively decided to follow do not seem to be the best ones; the important thing is that everyone is learning to have values-related discussions and become part of a household with a purpose beyond itself.

The next group-level includes workplaces, schools, sports clubs – any organisation to which people belong. Most of these groups have formal forums in which you would be able to introduce the idea of integrating carbon-saving practices. All schools have boards and all boards are required to have goals and strategic plans. Many government agencies and companies have staff or management meetings that allow any one to raise issues that concern them. Increasingly, there are competitions and funding to encourage groups to come up with creative methods of climate change mitigation. The key is to find a way to start the conversation within the group, and then to build an organisational identity project around being carbon-conscious.

Collective projects along these lines not only build meaning, but also build belonging – another critical element in human wellbeing. People who are embedded in positive relationships with others are healthier and happier[35] and projects such as counteracting climate change offer plenty of potential to build positive relationships within organisations. Especially in workplaces, where there maybe a competitive hidden agenda between staff, building a culture of sustainability can reduce it as people unite in a genuine team effort.

Projects at the organisation level also have immediate potential to change the social conspiracy that keeps things running exactly as they are. If sporting bodies decided that games for young players could involve just one or two local clubs then it would be plausible for many families to leave the car at home on a Saturday morning and walk to most games. If a retirement village resolved to provide only locally grown fruit and vegetables at meals, then residents would not be obliged to eat a Californian orange or buy their own locally-grown produce. If employees received free bus passes, good cycling facilities and car parks only for those who carpooled, there would be an immediate change in people's beliefs about the expected way to get to work. If schools chose not to participate in shopping-related fundraising, then parents would have less reason to jump in the car and buy products from the mall that keep coal-fired power plants running. All these things are feasible and can create in members a deep-felt identification with the organisation. We are taking a stand, they say. We are leaving a legacy for future generations.

Towns, cities, provinces and countries can also draw on people's sense of citizenship, their craving to take pride in the places with which they identify. People have died for New Zealand – why not reduce carbon emissions for New Zealand? To even begin to set the ground-work for New Zealanders to adopt this project, we would need to stop focusing on New Zealand's small overall contribution to greenhouse gas emissions and start focusing on what it would be like to lead the world in climate change action. Political parties would need to reach agreement on immediate priorities – perhaps deciding on three goals. Good contenders may include: reducing the number of barrels of oil imported or tonnes of rubbish going to landfill (by a set amount), or increasing the number of hectares of native forests planted, cycle trips into downtown areas, schools with climate-change action plans or bus passengers (again, by a set amount). The nation's progress on these goals could be published monthly in the daily newspapers and broadcast on TV, radio and the Internet. This could be accompanied with one or two stories behind the numbers, to give our progress a human face. Cities and towns could have a similar strategy – a few targets and well-publicised outcomes. It would be even easier for people to see their efforts impacting on the project at the town or city level, as not only will the numbers change, but there will also be visible changes in community life.

Last, but not least, is the possibility of drawing together people all over the world as a huge human group, committed to the wellbeing of the planet we share. How cool it would be to have a giant carbon-concentration display panel located in every major city in the world. I can almost see scenes like those broadcast at the beginning of the millennium, of people everywhere cheering, in this case, when the carbon concentration in the atmosphere goes down.

Tackling climate change is a psychological challenge, but it also has potential to create a new sense of purpose and belonging, for individuals, households, cities, countries and maybe even for humanity. It may well be the most important task facing our generation. Let's go for it.

ENDNOTES
1   Slovic, P. 1987. The Perception of Risk. *Science 236 (4799)*, 280-285.
2   Stern, N. 2007. *Stern Review of the Economics of Climate Change*. Cambridge, UK: Cambridge University Press, http://www.hm-treasury.gov.uk/Independent_Reviews/stern_review_economics_climate_change/sternreview_index.cfm (accessed 11 Sept 2007).
3   Petit et al. 1999. Climate and atmospheric history of the past 420,000 years from the Vostok ice core, Antarctica. *Nature 399*, 429–436.
4   Intergovernmental Panel on Climate Change. 2001 *Summary for Policymakers to Climate Change 2001*. Geneva: IPCC Secretariat.
5   See, for example, Arendt, A. et al. 2002. Rapid wastage of Alaska glaciers and their contribution to rising sea level. *Science 297*, 382–386. Rignot, E.J. 1998. Fast recession of a west Antarctic glacier. *Science 281 (5376)*, 549–551.
6   Church, J.A. and White, N.J. 2006. A 20th century acceleration in global sea-level rise. *Geophysical Research Letters 33*, L01602.
7   Webster, P.J., Holland, G.J., Curry, J.A., Chang, H.-R. 2005. Changes in tropical cyclone number, duration, and intensity in a warming environment. *Science 309*, 1844–1846.
8   Hughes, T.P. et al. 2003. Climate change, human impacts, and the resilience of coral reefs. *Science 301*, 929–933.
9   National Snow and Ice Data Center. 2005. Sea Ice Decline Intensifies. Press release for 28 September 2005. http://nsidc.org/news/press/20050928_trendscontinue.html (accessed 11 Sept 2007).
10 Cowling, S.A. et al. 2004. Contrasting simulated past and future responses of the Amazonian forest to atmospheric change. *Philosophical Transactions of the Royal Society 359*, 539–547.
11 Stainforth, D.A. et al. 2005. Uncertainty in predictions of the climate response to rising levels of greenhouse gases. *Nature 433*, 403–406.
12 http://nationalacademies.org/onpi/06072005.pdf (accessed 11 Sept 2007).
13 New Zealand Ministry for the Environment. 2006. *Understanding Climate Change – Get a Grasp of the Facts*. Wellington: New Zealand Ministry for the Environment.
14 2003 figure – Greenhouse Gas Emissions Data for 1990 – 2003 submitted to the United Nations Framework Convention on Climate Change, Key GHG Data, FAOSTAT (available at http://globalis.gvu.unu.edu/).
15 Monbiot, G. 2006. *Heat. How to Stop the Planet Burning*. London: Penguin.
16 http://www.mfe.govt.nz/publications/ser/gentle-footprints-may06/html/page7.html (accessed 11 Sept 2007).

17 https://myclimate.myclimate.org/ (accessed 11 Sept 2007).
18 Hillman, M. 2004. *How We Can Save the Planet*. London: Penguin.
19 See, for example, Doll, J. and Ajzen, I. 1992. Accessibility and stability of predictors in the theory of planned behaviour. *Journal of Personality and Social Psychology 63*, 754–765.
20 Geldof, B. & Vallely, P. 1986. *Is that it?* London: Macmillan
21 Latané, B., and Darley, J.M. 1976. Help in a crisis: bystander response to an emergency. In J.W. Thibaut and J.T. Spence (Eds). *Contemporary Topics in Social Psychology* (pp. 309–332). Morristown, NJ: General Learning Press.
22 Cialdini, R.B. & Trost, M.R. 1998. Social influence, social norms, conformity and compliance. In D.T. Gilbert, S.T. Fiske & G. Lindzey (Eds). *The Handbook of Social Psychology, Vol 2, 4th Edition*. Boston: McGraw-Hill.
23 See Piliavin, J.A. and Charng, H.-W. 1990. Altruism: a review of recent theory and research. *Annual Review of Sociology 16*, 27–65.
24 See Hogg, M.A. & Abrams, D. 1988. *Social identifications: A social psychology of intergroup relations and group processes*. London: Routledge.
25 Krushner, T. 2004. Despair is a lie we tell ourselves. In Loeb, P.R. (Ed.). *The Impossible Will Take a Little While* (pp. 169–174). New York: Basic Books.
26 *An Inconvenient Truth* (motion picture). 2006. Paramount Classics and Participant Productions.
27 http://www.makepovertyhistory.org.nz (accessed 12 Sept 2007)
28 http://www.one.org (accessed 12 Sept 2007).
29 Rachlin, H. and Green, L. 1972. Commitment, choice and self-control. *Journal of the Experimental Analysis of Behaviour 17*, 15–22.
30 Harré, N. 2007. Community service or activism as an identity project for youth. *Journal of Community Psychology 35*. 711-724.
31 Ryff, C.D. and Singer, B. 1998. The contours of positive human health. *Psychological Inquiry 9(1)*, 1–28.
32 Armstrong, K. 2004. *The Spiral Staircase*. New York: Arrow. p. 298
33 Thanks to Sonja Tepavac, Maree O'Neill and Donna Watson.
34 Covey, S. 1997. *The Seven Habits of Highly Effective Families*. Golden Books.
35 Baumeister, R.F. and Leary, M.R. 1995. The need to belong: desire for interpersonal attachments as a fundamental human motivation. *Psychological Bulletin 117*, 497–529.

Schools in a carbon neutral world

Heidi Mardon

With a background in sustainable architecture and environmental education, Heidi Mardon is National Director of The Enviroschools Programme, a large network of passionate people committed to Education for Sustainability. By working collaboratively, pooling skills and sharing knowledge, The Enviroschools Foundation supports students to design and create healthy, peaceful and sustainable schools and communities. The inspiring work of students, teachers, facilitators and supporters are showing New Zealand the possibilities for a sustainable future – through their words and actions today.

THE CURRENT SITUATION
There are growing numbers of schools and kura[1] working towards becoming more sustainable. However, the value of the school system as a catalyst for change is still being hugely underestimated. With increased recognition and a well-thought out support structure, schools could be leaders in their communities for sustainable, carbon-neutral living.

THE VISION FOR 2020
In 2020, schools will be models of sustainable living with buildings and grounds that reflect the culture and ecology of the community. Students will be making decisions and taking action in a large range of environmental and sustainability projects within the school and in the community. Teachers will be using student-centred approaches that stimulate enquiry, experimentation and collaboration. The whole learning process will be grounded in an ethic of care and creativity and schools will be centres for community learning.

STRATEGIES TO ACHIEVE THE VISION
Each school would develop a sound Sustainable School framework that integrates environmental education into all areas of school life and encourages sustainable practices. Professional development to support environmental education and sustainable schools must be a core part of teacher training and available for all school staff and management. Peer support networks would be available for students, teachers and schools for the sharing of experiences, resources and expertise. Schools would become the centre of community networks that link people and groups with expertise and passion for different areas of sustainability.

In Aotearoa/New Zealand we already have many of the people and processes in place to action such a vision. What is needed now is a strong collaborative approach between government and community agencies – working together for the well-being of people and nature.

Imagine learning from a school just by walking past it. It's raining and you see the rainwater being channelled off the roof into storage tanks and wetlands – or if it's sunny a digital map shows how many buildings are receiving electricity from their solar panels. Carved gates tell the story of the land and who has been before, and through the gates you can see high tech solar equipment sitting alongside student-made contraptions for collecting sun and water. The students are growing food, composting, building structures and going about myriad environment-sustaining activities that you could do yourself at home. As well as simply looking at the school as you walk past, you can go to its open days. At the open days, students show you around – they have become the teachers, explaining the different systems that they have designed and implemented. They draw your attention to a huge plan, which shows numerous other projects being designed for future action within the school and also projects that the school and community are working on together – the zero waste system set up at the neighbourhood shops and the local stream restoration. The students explain how each project works and why it is needed.

This chapter invites you to consider the possibility that school is one of the most powerful institutions we have for creating a carbon-neutral country – and that the way forward is by creating sustainable schools. These are schools that have a culture of environmental and community care, where a reduction in carbon emissions is simply one benefit alongside many other aspects of community and environmental well-being.

Creating more sustainable ways of living requires different ways of thinking. For this we will need the expertise of environmentally-aware architects, 'green' accountants, sustainable farmers, carbon-counting engineers and the commitment of conscious consumers. We need the wisdom of our elders and the lessons learnt by our ancestors – and we need to envisage what environmental health and sustainable communities actually look like. To create this new and exciting place requires a different kind of decision-making. Therefore, we need schooling that will help us develop ourselves as communities of people who:

- Are knowledgeable workers who do our jobs in a carbon-neutral, sustainable way
- Want to be active citizens who get involved in decision-making about the health and well-being of our place
- Know our whakapapa and heritage so that old wisdom can inform new practice
- Become reflective community members who want to care for *all* living things in our communities.

If schools empowered us to be these people what would the schools of 2020 look like and feel like? Would schools be any different from what we have now? What are the possibilities?

INSPIRATION, LEARNING AND EMPOWERMENT

Young people want wild places to play, they want to bring back the birds and they do not like 'nothing spaces'. They want litter-free school grounds, with fruit trees and peaceful places to sit. These are some of the themes that recur when students are asked to explore their school environment and say what they would change about it. So, what has this got to do with the school curriculum or creating a carbon neutral sustainable world? Everything!

Climate change is one symptom of unsustainable human behaviour – and it is a problem. But beginning with a problem – especially one that someone else has created – does not generally inspire young people (or anyone else) to change the way that they do things. Effective learning and change begins with what motivates young people. And focusing on learning that creates positive change instead of solving negative problems provides an abundance of learning possibilities and benefits for students, schools and their community. The powerful combination of student motivation and focusing on positive change has already borne fruit at a school in Hamilton. The desire for wild places in which to play provided Hukanui school students with the motivation to plant hundreds of trees in a derelict gully behind their school in Kirikiriroa, Hamilton. Students designed educational maps to show others how valuable the gully was, and they sought expert community help, got funding and designed areas for planting and for play. The project initiated a whole-school environmental journey that has now been going on for eight years and involves neighbours and the wider community. Students give guided tours through the beautiful revegetating wilderness, and participate in the design and landscaping of other parts of the school grounds. They have also developed a worm farm and native plant nursery, to maintain the long-term vision of the gully project. The students' latest venture is the design of an 'enviro-classroom'.

The Hukanui project is just one example of how focusing on creating a special place, rather than solving a negative problem, creates an abundance of benefits. If you look at it from a climate change perspective, you see the benefit of carbon emissions being offset by the increased number of trees growing. If you examine it from a biodiversity perspective, you find that increasing numbers of species of birds and insects are coming back into the city. From a teacher's perspective, students are gaining skills, knowledge and experience in a range of curriculum areas. And when parents are asked about the benefits of their child's involvement, we hear stories of increased confidence and excitement about learning. Schools also report reduced bullying and vandalism through students working together and feeling ownership of their surroundings. And from a student's perspective, it really is fun!

ENVIROSCHOOLS: EMPOWERING STUDENTS TO CREATE POSITIVE CHANGE

One of the amazing opportunities that a school provides for creating sustainability is that it has all the characteristics of a community, city or town but with the sole purpose of enabling learning. The school occupies a physical place, has buildings and infrastructure, resource flows, a political structure and a management system through which decisions are made. People of all ages occupy this world to learn and teach for six or more hours a day. What better way to create a country of carbon-neutral citizens than to actively be one of those citizens at school, and work with your community to do it?

In the early 1990s, the seeds of this vision were present in the minds of a few creative people, who later initiated the Enviroschools Programme. Enviroschools was a positive response to the increasingly doom-and-gloom nature of New Zealand's environmental situation. Through experimentation, wananga and collaboration between people with a diversity of viewpoints, a journey was begun. The journey now involves over 500 schools and kura – of all age levels, sizes and deciles – and a large number of local authorities, educators and community members creating and reflecting on environmental education and sustainable schools.

The Enviroschools mission (kaupapa) is the empowerment of young people to create a healthy, peaceful, sustainable world, and the Programme has five Principles that underpin this whole-school approach:[2]

1. Genuine student participation comes by listening to students' own unique and creative perspective. Including young people in decision-making and action empowers them to be active environmental citizens for life and enriches the development of the whole school environment.
2. Environmental Education is an action-focused approach to learning that engages students in the physical, social, cultural and political aspects of their environment.

3. Maori perspectives and knowledge of the environment offer insights unique to the culture with the longest history of human interaction with this country. Including Maori perspectives enriches the learning process and honours the status of indigenous people in this land.
4. Respect for the diversity of people and cultures is integral to achieving a sustainable environment in New Zealand – one that is fair, peaceful and co-operative and that makes the most of our rich cultural traditions.
5. Sustainable communities act in ways that nurture people and nature, now and in the future.

These five Principles are applied across four key areas of the school – place, people, practices and programmes.

Place: healthy physical surroundings, whenua, ecological buildings
In a sustainable school the design of buildings and landscapes teaches students about their connection to the land and their community – a sustainable school reflects the culture and heritage of its place and is designed from ecological and participatory principles. A sustainable school enables students to design their places and have access to see how buildings produce energy, save water, nourish the landscape and are healthy for all living things. The school grounds demonstrate how ecosystems work and provide students with opportunities for experiencing their interconnection with nature.

People: participatory management, whakapapa, whanaungatanga
In a sustainable school the decision-making is done with the involvement of students and other members of the community. Sustainable schools establish a student Envirogroup that represents classes and steers a whole-school sustainability plan. The school has a whole-school vision for sustainability and there are regular meetings between students and the board of trustees to develop policies and plan budgets. Whanau and community members are invited to participate in major decisions.

Practices: sustainable operating practices, tikanga
A sustainable school operates on sustainability and conservation principles, such as energy efficiency, waste minimisation and water conservation. Green purchasing policies are also adopted, encouraging school-bought products to be environmentally and ethically sound.

A sustainable school will also develop Care Codes and incorporate practices to promote and lead to sustainable lifestyles. Developing these systems is a learning opportunity and creates an environmentally aware culture within the school. All school members have a role in maintaining these practices.

Programmes: a living curriculum, student-centred learning, ako
A sustainable school enables students to gain skills, knowledge and competencies through a range of activities and experiences that enrich the formal curriculum. Students learn about sustainability through creating a sustainable school and community. This involves students in: designing and implementing sustainable projects, learning from role models in the school, initiating their own learning through play and tapping into the wisdom of the wider community. Learning also involves teaching – a living curriculum will enable students to share their learning and become mentors and leaders for other schools and groups in the community.

Being an Enviroschool is a journey of exploration, experimentation and sharing different ways of seeing. We cannot get sustainable schools by simply employing people to make them, they are created through learning and action.

PUTTING THESE PRINCIPLES INTO PRACTICE: CREATING A LIVING CURRICULUM
If you ask young people about environmental and community issues, you will hear many of them say that they do not want to be seen as having 'potential to contribute in the future', they want to been seen as contributors now. Schools and the education system can support students of all ages to contribute to a sustainable world through an action-based approach to learning. In an Enviroschool students are creating their own sustainable school by being involved at a range of levels - from the design of a classroom rainwater collector to the implementation, maintenance and monitoring of a whole-school zero waste system.

A living curriculum enables students to work in partnership with a range of different people. One of the keys to empowering students to create sustainable schools is allowing them access to the diversity of wisdom in their school and wider community. In this way students can learn collaboratively, building knowledge from many different sources and perspectives, each playing a role in the development of ideas and projects. Group learning and action reflect what is needed in the wider world, where issues of sustainability cannot be resolved by individuals alone; instead, the solutions need a variety of different skills and input.

Through an action-based learning approach students are supported to identify issues that are important to them and to explore alternative solutions. They do this by observing their own environment, talking with people in the community and researching local and global topics. They look at how the situation came to be the way it is, and what it was like before. They also express their own knowledge and viewpoints and share ideas collaboratively – pooling knowledge for all to use. Based on the knowledge they gain, students make decisions, which involves exploring a range of alternatives and asking critical questions such as: What criteria will produce the most sustainable options? Who gains and who loses from these decisions? Students then design and plan the actions and projects that they want to undertake. Design and decision-making engages students in a suite of skills and competencies that are necessary for creating sustainability.

Students are given responsibility for monitoring and maintaining their own projects, since the maintenance of any project is just as important as its initial design and set-up. Enabling students to access the more operational aspects of school life is crucial if they are to have the ongoing learning experience of monitoring and looking after the system or project that they have established. Students are also encouraged to pass on their knowledge or roles as sustainable school 'managers', ensuring that the system is maintained as students move through and leave the school.

Finally, students reflect on and celebrate their personal and collective achievements. Students need time to think about how things went, what has changed and what they would do differently. It is this reflective aspect of learning that provides the opportunity for students to be aware of what they have learnt, how they learn and how their attitudes and values may change.

The project being undertaken by students from Te Kura Kaupapa Maori o Te Rawhitiroa in Te Tai Tokerau, Northland is to restore the mauri (life force) of their awa (river). The project is an example of a holistic approach to action-based learning. The teachers supported the students' learning journey by asking them how they wanted to do it and who they wanted to help them. Students began by walking the banks of the awa – and then asked a local helicopter pilot to fly them the length of the river so that they could see the whole state of its health. They sought people who knew the river many years ago and could tell them what it was like when it was healthy. Students then created presentations for community radio and organised clean-ups and tree plantings, and they continue to raise local awareness of the needs of the river. Through this project, the kura students have inspired community action and long-term commitment to the care of their environment. They have done this in partnership with teachers and other adults who have gone on the learning journey with them and supported them to become empowered through taking action about something meaningful in their local community.

COMMUNITY SUPPORT FOR SUSTAINABLE SCHOOLS
If schools are to use the formal curriculum as an opportunity for action-based learning and empower students to design and implement their own projects, then teachers and all people who work in schools need support in developing Education for Sustainability.

A number of programme providers collaborate to extend these possibilities for students and teaching staff in schools. Alongside the sustainable school approach of Enviroschools two other national support systems are available to schools – The National Education for Sustainability (NEfS) programme and the Matauranga Taiao programme for kura – both led by the Ministry of Education. The NEfS programme operates through the school advisory service in the six University Schools of Education and NEfS advisors are available to work with teachers to integrate sustainability into the curriculum through one-on-one planning, staff meetings and regional workshops. Matauranga Taiao began this year and is currently establishing a support network through professional development workshops.[3]

There are also many effective programmes run by local authorities about local environmental issues such as waste and water, and a range of community-driven initiatives. These programmes work together to support a holistic process in schools. Professional support for boards of trustees and school management would also be valuable to enable managers to be creative in terms of the sustainable management of budgets and school operations.

The architects and planners who design our schools also have a powerful impact on sustainability and student learning through the physical environment that they create. Many students and schools are currently leading the sustainable development of their local landscapes and waste systems, and there is a growing demand by schools for the built environment to also support students' environmental curriculum. Schools need buildings that involve, engage, inspire and teach students and the community about a sustainable world. Sustainable schools need experts in the construction industry who are prepared to learn alongside them and to evolve and reshape our built environment so that it nurtures people and nature.

Finally, although Enviroschools operates to support the teachers and students at the school level, there is a real need for support at the tertiary level. In a country aiming to be carbon neutral, core courses in sustainability should be mandatory in all tertiary programmes. In particular, teachers, school administrators and school designers have key roles in the development of sustainable schools and communities, and this should be reflected in the educational programmes they have access to at the tertiary level.

CONNECTING CARBON TO OUR LIVES
Enviroschools now has over a decade of experience, developed through the considerable contributions of thousands of people of all ages. The lessons that we have learnt along the way are many, but in relation to the issue of carbon neutrality they come down to one big issue – that of connectivity. Whilst there may be little doubt that we need to reduce our CO_2 emissions, we should remember that enhanced climate change is just one symptom of the bigger issue of unsustainable living, which is about people and the communities in which we all live in. If we are to reduce carbon emissions as part of a more sustainable lifestyle, we need to see the links to the fundamentals needed for life.

Manaaki tangata, caring for people, is at the core of a sustainable school and a sustainable community. In the process of developing Enviroschools, we have seen the power of trust, love and reciprocity to support people to create change in their own lives and communities. Schools have huge potential to bring communities together to care for and support each other.

If we had set about telling schools exactly what they must teach in order to reduce carbon emissions, rather than provide them with guiding principles, it is doubtful that we would be seeing the energetic enthusiasm that we observe in students and teachers around the country. Creative exploration, questioning, design and experimentation energises us to take action.

Learning can be a joyous endeavour that inspires us to live creatively and sustainably together. If student empowerment and ecological principles underpinned all of our learning from a young age, then carbon-neutral behaviour would simply be one aspect of how we choose to live. Celebrating the combined wealth that we have in our schools and communities – the wealth of skills, talents and knowledge – reduces our need to 'buy' our satisfaction in other forms.

Around one-third of New Zealanders are children and young people. This sector of our community can make a significant difference right now if it is actively engaged in decision-making and action. Creating a sustainable school is an ongoing learning process that can begin with small actions and lead to whole-school and community change. In this country we have a wealth of experience in environmental education and dedicated networks of passionate people developing programmes and supporting students – the sustainable schools journey has already begun. If sustainable schools became centres of learning for their communities we could all learn together about how to create a sustainable Aotearoa/New Zealand.

ENDNOTES
1   Maori schools
2   For more information about the programme and its principles, visit http://www.enviroschools.org.nz
3   More information can be found by contacting the New Zealand Ministry of Education or http://www.tki.org.nz (accessed 1 Oct 2007).

A new paradigm for home renovation

Jette de Jong and Niel de Jong;
Research Assistance by Joanne Nikolaou

Niel and Jette de Jong are the Directors of Heritage Design Group, an Auckland-based architecture practice with a philosophy of creating 'architecture for your future'. This focus on sustainability is applied equally to renovations and new homes, taking into account the life cycle of the building and how a home can positively influence its inhabitants to live a sustainable and carbon-neutral lifestyle. Strong proponents of 'walking the talk', the property where the de Jongs live and work is a heritage home that they are working towards making 100 per cent sustainable through renovation and the application of technology.

THE CURRENT SITUATION

When trying to create more space and make their homes appear more fashionable, many New Zealanders approach renovation as a cheaper option than buying or building a new house. Focusing on getting things done for the lowest possible up-front cost, they will include carbon-saving measures only if these are required by law. Issues such as sustainability and the toxicity of materials are not even considered. This often results in sub-standard, piecemeal renovations that fail to address fundamental issues – and in a few years' time more renovations are required as the owners try to fix the same problems or, even worse, additional problems caused by the previous work. In the end, all that is achieved is wasted effort and materials, and a home that is unpleasant and frustrating to live in.

THE VISION FOR 2020

In 2020, renovation will be seen as a preferred option for building as it involves reusing and recycling materials that have already paid their debt to the environment. Renovations will be planned carefully to make the best use of the money and resources invested in the structure – they will solve existing design flaws and inefficiencies and provide for the future needs of the occupants. And they will also avoid adding new problems such as poor layout, inflexible design, and toxic or unsustainable materials.

People buying, selling and living in houses, as well as those who design and build them, will understand and value the importance of keeping the carbon footprint of a home to a minimum. Factors such as passive heating and cooling, designs that encourage low resource consumption, long-life materials, and high-tech options to improve energy efficiency will be highly valued by buyers and renters. Homes that include these features will command premium prices.

STRATEGIES TO ACHIEVE THE VISION

Knowledge is power, so a key strategy will be to ensure that consumers and all those involved in the building and housing industries have the information they need to make informed decisions about buying, selling, building and renovating homes.

Where the provision of information and education does not result in the changes in behaviour required to become carbon neutral by 2020, local and/or nationwide regulations will need to be put in place to support this goal.

INTRODUCTION

When we tell people that our architecture practice specialises in heritage renovations and creating sustainable homes, they often find it difficult to contemplate the possibility of combining these two disciplines. The concept of a carbon-neutral bungalow seems like a paradox. The challenge of making a villa practical for 21st century living seems big enough, without trying to turn it into an 'eco-house' as well. When we raise the idea of sustainability with clients who are thinking of renovating, questions abound.

> 'If we want a carbon-neutral home, wouldn't we be better off knocking down the existing house and building a new one from scratch?'
> 'If I invest in design and technology to make my home carbon neutral, what's the payback?'
> 'What does "sustainable" mean?'
> 'What exactly is meant by the "carbon footprint" of my house, anyway?'

In answering these questions, this chapter provides practical ideas for overcoming the many challenges around renovating to create a carbon-neutral home. Some of these are physical; some are attitudinal; many are a bit of both. Reading and understanding this chapter will help you to see your home as a living organism that has a life cycle of its own, which both influences and is influenced by its inhabitants and the wider community. This chapter addresses a number of widely held preconceptions about what renovation should entail and how it should be done, and presents some alternative models for working with your home. Finally, this chapter provides a vision for how individuals, groups, organisations and governments will think and behave in the future if we are to achieve our goal of carbon-neutral renovations by 2020.

The carbon footprint of your home

The Introduction to this book has provided a general overview of the carbon issue, and what types of activities produce CO_2, but how does this relate specifically to our homes? One succinct definition of a carbon footprint that can be easily related to a building is 'a measure of the exclusive total amount of carbon dioxide emissions that is directly and indirectly caused by an activity or is accumulated over the life stages of a product.'[1]

Your home's carbon footprint is derived from both essential elements of the above definition:

- *Life stages of a product* relates to the materials used in the building itself (for example, the timber, steel, concrete, aluminium, glass, linings, insulation, etc.) – including the energy and resources used to produce and transport them; how long the materials last; what maintenance they require; and what happens to them at the end of their life.
- *Activity* relates to the way we live in our houses – including the energy and resources needed for heating and cooling, lighting, cooking, cleaning and other functions; how waste is managed and disposed of; how the home's location and amenities allow the occupants to interface with the community, especially with regard to travelling to and from work, shops, schools and other important places.

Sustainable building

We believe that the carbon footprint of a home or building is just part of how it affects the environment, so we have also addressed a number of areas relating to the wider issue of sustainability in this chapter. The New Zealand Ministry for the Environment (MFE) defines sustainable buildings as 'sensitive to:

- The environment – local and global
- Resource, water and energy consumption
- The quality of the [indoor] environment – its impact on occupants
- Financial impact – cost-effectiveness from a long-term, full financial cost-return point of view
- Long-term energy efficiency over the life of the building.' [2]

THE CHALLENGE: CAN WE DO IT? YES WE CAN!

So, is renovation really a practical option for achieving a carbon-neutral, sustainable home? Is it really as good as (or even better than) demolishing a tired old house and starting again? And does it make sense financially?

We believe it is not only possible, but in fact it is usually preferable to work with what we already have to create the sustainable homes of our future. This view is backed up by MFE, which claims that to 'renovate an existing building ... is the most sustainable construction option'.[3] However, MFE also states it is 'a common myth that it is possible to make a building sustainable simply by adding some energy-efficiency and water conservation measures'.[2]

Let us take a look, then, at how we can renovate our homes in a way that is more than skin deep. Let us think about how we can overcome physical, lifestyle and technological challenges, as well as change our attitudes and thinking, to arrive at a new paradigm for home renovation by 2020 – and a new model for the prestige home of the future.

Grand planning

The most important factor in a successful home renovation is not usually how it looks at the end, how much it costs or even how quickly and smoothly the work progresses (although these are all important). It is having a 'grand plan'. Rather than trying to address your home's problems one at a time – endlessly changing bits and pieces and ending up with a grand mess – a grand plan allows you to address each problem once and fix it properly, with the minimum cost and use of resources.

For example, if you need an extra bedroom, and also need to renovate a kitchen or bathroom, but your lounge is always cold at night while the kitchen overheats in the afternoon sun, it makes sense to decide on the best location for all of these rooms before investing any money or resources in building.

We have seen many homes where the current or previous owners have enthusiastically tackled a small project because they could not afford more major work, but the 'quick fix' has actually increased the costs - and restricted the options – for solving other issues (see Case study: Curse of the spiral staircase).

If budget is a major factor, good planning should actually reduce renovation costs in the long term. Once you have an overall plan, you can have the proposed work costed and compare it with your budget. If you can't afford to do all the work at once, breaking down the work into smaller jobs will allow you to proceed in stages as money becomes available – without compromising on the final result.

Employing an architect or designer who uses three-dimensional (3D) computerised Building Information Modelling (BIM) tools can add a lot of value to renovation planning. A 3D computer model allows you to virtually 'walk through' the changes proposed to your home and better understand the sizes of and relationships between existing and new spaces, as well as allowing you to easily ensure that there will be the right amount of daylight, warmth and shading when you want it. The BIM approach also makes it easy to clarify economic and logical ways of breaking a project into stages if required.

> **CASE STUDY**
> **Curse of the spiral staircase**
> We were asked to look at a very unusual circular home that had been self-built by the original owner in the 1940s, using many found and recycled materials. The current owners needed more space, and had been puzzling for some time about how they could achieve this without destroying the home's quirky charm.
> As we walked through the house discussing options, we realised that many of our suggestions and ideas were receiving a fairly lukewarm response. After some more conversation, the reason for this became apparent – just a few months earlier, the man of the house had gone to considerable trouble and effort to install a spiral staircase ('donated' by a friend) himself, and he was very reluctant to have any of his work undone or changed.
> Unfortunately, the staircase was, in our view, located in a very unsuitable position and retaining it severely restricted the design options available to address the family's other needs. In addition, it did not meet current Building Code requirements and the work had been done without a building permit.
> If there had been a grand plan for this home before any renovations had begun, the staircase project would never have got off the ground. There would have been no materials and labour wasted in installing it, and there would have been no impediments (physical or emotional) to making other improvements that would really work for the home's occupants.

Keeping warm and cool

Heating and cooling are two of the greatest consumers of energy (and therefore producers of carbon) in our homes, so reducing the need to burn fossil fuels or use electricity to do either of these effectively is probably the most important way in which we can reduce the carbon footprint of our homes.

When starting from scratch, it seems easy enough to get the basics right – good sun orientation, proper insulation and thermal massing for heat retention in winter, and carefully planned overhangs and ventilation to provide shade and cooling cross-breezes in summer. But when we have an 80-year-old bungalow with all the living areas on the south side, how do we even begin to get the temperature regulating systems working sustainably for the future?

Passive heating Passive heating is an inexpensive and logical way of reducing a home's carbon footprint, and should be one of the first considerations when developing a grand plan. If it is at all practical to re-orient living areas towards the north as part of a renovation, relegating service areas (such as bathrooms, laundries, garaging and storage) to the south side, you should do this. The sun is the most powerful source of heating and light we have for our homes, and it is free – so it makes sense to use this energy in the areas where we live and sleep. Spending an extra 10-20 per cent of your renovation budget on improving your home's orientation to the sun will pay long-term dividends both in the energy efficiency of your home, and in your comfort and enjoyment living there. It will also greatly increase the value you get from the rest of the money you spend on the renovation.

If the current design of your home and/or its position on the land make it impossible to allow direct sun into one or more critical areas, combining solar water heating with solar under-floor or radiator systems is another way of using the sun's energy to help heat your home. Solar under-floor heating is a particularly good option when developing basement areas, as it can be installed easily at the time a new concrete floor slab is laid. Radiators containing solar-heated water can be used instead of electrically heated towel rails in bathrooms, or to provide ambient heat in bedrooms and living areas.

Existing fireplaces and chimneys should not be dismissed as a useful way to improve heating. While open fires are obviously not efficient or environmentally friendly in terms of emissions, many old villas and (to a lesser degree) bungalows feature large brick chimney structures in the centre of the house, which originally had a coal range on one side and fireplace on the other. In the past, the coal range was run every day for cooking and the entire chimney structure would warm up, slowly releasing this heat into the house overnight. Using an existing brick chimney or creating a solid concrete mass to house a new energy-efficient wood-burning firebox is a sensible way to revive this thermal massing model – adding a wet back to boost solar water heating increases the benefits further.

Passive cooling and ventilation In addition to heating, your grand plan should also make provision for how you will deal with unwanted heat – and moisture – an important, but often overlooked, consideration in New Zealand's sub-tropical climate. The wide eaves and verandas of our traditional bungalows and villas are often well proportioned for keeping out the fiercest summer sun, and the small, top-hinged 'fan light' windows popular in bungalow and state house joinery until the 1950s allow these homes to be well ventilated, even in rainy weather. Louvre windows perform a similar function in many homes built in the 1960s and '70s. It is critical to maintain and enhance these benefits when renovating by paying careful attention to new and altered rooflines, and ensuring that new joinery includes at least some smaller windows that can be left open in all weathers. Ideally, these will match or complement those in the existing house.

Where it is not possible to obtain good shading and ventilation by matching the original roofline and window design of your home, ventilating skylights, modern complementary louvres and wind-out awnings can often be incorporated sympathetically into a renovation. The addition of a double-glazed staircase on the northeast or northwest side of the house can also provide an ideal opportunity to add passive heating and cooling to a home. During the winter, if you keep the windows in the stairwell closed and shut the doors to the ground floor or basement, heat will be collected and dispersed to the upper levels of your home. In the summer, if you open windows at the top of the staircase and on the upper levels of the house, and also leave the doors at the bottom of the staircase open, the hottest air will escape, pulling cool air through the house from the lower levels.

Insulation While a certain level of insulation is required in all new homes and renovations in New Zealand, the general rule is, more is usually better. So, if your ceiling and wall cavities allow for more than the minimum level of insulation required, it is well worth spending just a little bit extra to get a higher R value[4]. Bear in mind though that R values indicate the levels an insulation product is capable of achieving if it is installed correctly. Large gaps left in difficult to reach areas or around recessed lighting (see 'Taking an enlightened approach to lighting' later in this chapter) can render even the best insulation virtually useless, so, whether putting it in yourself or employing professionals, ensure that the insulation material is fitted snugly into every space.

If you choose an insulating material that performs another function as well, you will get even more value out of it. For example, many people install fibreglass batts in walls and ceilings, but research has shown that 'wool [insulation] absorbs and retains indoor air pollutants such as sulphur dioxide and nitrogen dioxide (which comes from gas stoves and heaters), and formaldehyde which is emitted from common building materials.'[5] In fact, 'tests have shown that a house with wool insulation is able to absorb up to 1000 times the usual amount of formaldehyde present in that house.'[6] Because wool retains these pollutants, not releasing them into the air again, using wool to insulate your home makes it safer as well as warmer to live in.

Under floors, batts or foil alone can be somewhat ineffective as both will admit a certain amount of draught and cold air from below. While batts and foil can be used in combination to keep warm air in and stop cold air from entering through the floor, polystyrene is a more effective – and cost-effective – option for doing this. Despite popular opinion, polystyrene is also a good choice environmentally when it is made from recycled or waste materials using a low energy CFC-free process and the manufacturer will accept it back for further recycling at the end of the building's life. At least one company in New Zealand currently meets all of these criteria, so it is worth checking with the supplier before selecting a brand of polystyrene insulation.

When adding or upgrading under-floor insulation is difficult or impossible, you can still improve the indoor climate of your home through your choice of fittings. If there are areas of your home that have no or insufficient sub-floor space to allow the installation of under-floor insulation, avoid the fashion for polished wooden floors and install environmentally certified woollen or recycled carpet with a thick insulating underlay, or natural cork tiles, which have excellent thermal and environmental properties.

Glazing Glazing is an important, but often overlooked, component of a home's total insulation system – although it is likely to receive more attention when proposed regulations to make double glazing mandatory come into force in the near future. While double glazing is clearly the most energy-efficient choice for new joinery, installing it as part of a renovation raises a number of problems in terms of matching and upgrading existing windows, and these are discussed below.

One of the advantages of renovating over building from scratch is that certain materials, in particular joinery, can be reused. However, most of the traditional timber joinery profiles used in older New Zealand homes cannot have Insulating Glass Units (IGUs) simply retro-fitted – which means that new window sashes must be made to accommodate double glazing. This is likely to increase the cost of reusing existing timber joinery to a point where it becomes uneconomic, almost forcing us to install new windows and doors and relegating items that are still useful to the demolition heap.

Where regulations allow, and where the winter climate is not too severe, an alternative to completely new joinery is to replace the glass in existing joinery with a low emissivity or solar control glass, which is about one-third more effective than standard six-millimetre-thick clear glass at retaining heat inside the house in winter and resisting the penetration of the sun's heat in summer.[7] Adding heavy floor-length lined drapes is also an effective way to retain heat inside a house at night (although less useful for keeping it out during the day in summer!).

When new joinery is required for a renovation, it may be tempting to use aluminium doors and windows, as modern double glazing units are generally designed for these, and they are the most cost-effective and low-maintenance option. However, in our view, matching the style and materials of existing joinery is as important to your home's sustainability as it is for maintaining its aesthetic appeal (see 'Flexibility for the future' later in this chapter for more on this). For most homes built before the 1970s, this means sticking with timber joinery – whether it be recycled or manufactured to match – but there can be exceptions (see Case study: Timber or aluminium windows – a transparent choice).

CASE STUDY
Timber or aluminium windows – a transparent choice
Our clients, who were committed environmentalists, wanted to improve the energy efficiency of their 1960s board-and-batten home in Taupo as part of a larger renovation project. One of the key components of the brief was to double glaze the entire house, especially on the south side, which faced the lake.

While it is our normal practice to reuse joinery and other components whenever possible in a renovation, in this case most of the joinery was in poor condition, and only one large window was the right size to use in the new extension. Furthermore, the opening sashes would have to be remade to accommodate double glazing units. Rather than remaking the windows in timber, we elected to replace all of the existing joinery with composite (aluminium exterior/ timber interior) units for the following reasons:

- The existing joinery profile was a chunky 1960s style that could be easily replicated in aluminium without spoiling the 'kiwi bach' character of the house.
- Aluminium is a very low-maintenance material, making it a sensible choice for external joinery. While it has a high level of embodied energy (meaning considerable energy and non-renewable resources have gone into its production), this is compensated for by its long lifespan and ability to be recycled. However, if aluminium is also used inside, cold can be easily transferred through the metal surfaces. Many people also find this highly processed metal less appealing than natural timber to live with in their indoor environments.
- Timber is a renewable resource that is biodegradable at the end of its life, and, when finished with a natural oil or wax finish, provides a healthy, pleasant indoor environment. It also acts as a very effective thermal barrier, preventing the transfer of cold, outside air to the interior of a home. However, the need to repaint timber every five to eight years makes it a less sustainable option for exteriors, due to the labour and resources that painting consumes. And, while modern environmentally sustainable paint residues are less toxic than those used by previous generations,[8] we should still consider their potential effects on our environment.

Living neutrally

Before renovating, we are likely to have analysed the features of our existing home that are working and not working in relation to our current lifestyle and family culture, and have probably developed a 'wish list' of problems to be solved through renovating. However, something we may not have done is to consider how altering our home and the things in it may assist us to develop a more sustainable lifestyle – without major inconvenience or additional time and effort.

Service areas such as kitchens, bathrooms and laundries are the ideal target for this approach, as small behind-the-scenes changes in these areas have the potential to vastly reduce the amount of water and energy we consume, the amount of rubbish we produce, and how we dispose of this rubbish.

Checking the Star ratings as well as the style Appliances such as fridges, washing machines and dishwashers may not strictly form part of the fabric of a home, but they are often purchased as part of a building or renovation project and frequently influence the dimensions of a room, the design of built-in cabinetry and other more permanent fixtures. While the up-front cost of home appliances make buying them a relatively major purchasing decision in most households, their energy and (if applicable) water consumption has the potential to significantly add to – or reduce – this cost over time. However, a quick informal survey among some of our clients and friends suggests that, for many consumers, ongoing financial and environmental costs remain largely irrelevant in their appliance purchasing decisions.

Energy- and water-efficient appliances are readily available (although sadly those made in New Zealand lag far behind in this area; see Case study: New Zealand made or energy efficient?), so why are there not more people making a conscious decision to purchase them? We believe it is partly a consumer education issue, and partly an information availability one. New Zealand has recently adopted the international Energy Star system for evaluating the energy and water consumption of various appliances.[9] However, at the time of writing, there was no information or incomplete information available on the New Zealand website for comparing many types of appliances. The Australian Energy Rating[10] and Water Efficiency Labelling and Standards (WELS)[11] schemes are good sources of information until a full range of product ratings are available for New Zealand.

> **CASE STUDY**
> **New Zealand made or energy efficient?**
> As part of a recent project to improve the energy efficiency of our own home and offices, we decided to replace our ancient appliances – fridge, freezer, dishwasher and washing machine. We began our research with the best of intentions to buy New Zealand made, but were utterly dismayed to find that only a single fridge-freezer model produced by New Zealand's major whiteware manufacturer achieved the minimum four-star rating (with five stars being the best and one being the worst) we were seeking. The best energy ratings for New Zealand-made dishwashers and washing machines were only three stars, with water consumption equally poor (i.e., quite high) for both of these items.
>
> While many of the appliances with the best energy ratings and

low water consumption were the more expensive European brands, we found a number of models with good ratings that were available in a more modest price bracket as well. Our experience left us wondering why our local manufacturer does not put more effort into developing more water- and energy-efficient products? Who would buy a fridge with a two-star energy rating when there are much more efficient models available at a similar or lower price? And why are poorly-performing imported appliances even allowed to be sold in New Zealand?

The way fridges and freezers are installed is as important as the Star rating for energy efficiency. Most installation instructions recommend positioning the fridge away from sources of heat, such as windows and stoves, and leaving air gaps at the sides, top and back. However, it is often not clear that these guidelines are designed to ensure the appliance runs efficiently, and it is surprising how frequently they are not followed. A further practice (not usually mentioned by manufacturers) that can make a fridge run up to 25 per cent more efficiently is venting both the floor and ceiling at the back of the fridge cavity to the outside. This passive venting system pulls cool air from the sub-floor over the condenser mechanism at the back of the fridge, keeping it at a low, even temperature.[12]

Making water work In addition to selecting energy- and water-efficient appliances, one of the best ways to build sustainability into a home renovation is to locate the hot water cylinder as close as possible to its point of use.[13] This has the dual effect of minimising the amount of water wasted waiting for hot water to reach the tap, and minimising the amount of energy used to reheat the water displaced from the cylinder. If you use hot water in different parts of the house, you may even wish to consider multiple water heating options. For example, you could have: a large, solar heated cylinder to heat water for bathing, showering and laundry; a small electric cylinder under the sink for short, frequent bursts of hot water in the kitchen; and an electric cylinder or instantaneous water heating unit for hand washing in the bathroom.

Low-flow faucets, taps and showerheads can also be easily built into your design for kitchens, bathrooms and laundries, enabling you to conserve water without making any changes at all to your lifestyle. The WELS scheme[10] provides water-use ratings for tapware as well as appliances, although we have found it very difficult to find attractive-looking low-flow tapware in traditional styles to suit older period homes. Alternatives are: to fit an aerator to selected tapware (or use the existing faucet, which can be re-chromed if necessary); add flow restrictors to some or all of the taps in the house as part of your renovation; or install a pressure-limiting valve to your entire water supply.

If you are adding or replacing toilet cisterns as part of a renovation, it makes sense to choose a dual-flush option. However, if you want to use an existing or recycled cistern (for example, to match the period of your house), you can make it consume less water by adding a water-displacing mass such as bricks to reduce the amount of water flushed, or installing a 'gizmo'. This is a small lead weight with a hook on top of it, the weight of which will close the water outlet as soon as you release the flush mechanism, enabling you to control how long each flush lasts.[12] Adding a rainwater-collection tank and plumbing it to your toilet and other non-potable outlets is another way to make better use of our water resources without major changes to your lifestyle.

Kitchen consumption and waste The kitchen is the place in our homes where we generally consume the most, and create the most waste, all of which contribute to our carbon footprint, so it makes sense to build in solutions for dealing with this. Providing generous pantry and storage areas for food and other supplies will allow for a culture of bulk buying, reducing unnecessary packaging and minimising the need for trips to the supermarket. The only storage that should not be oversized are fridges and freezers, as larger models consume more energy than smaller ones, and all fridges and freezers run less efficiently when they are only half full of food.

When it comes to rubbish, at least three bins – one for organic material; one for recyclable plastics, metals and glass; and one for inorganic waste – are essential, as is a designated place to keep paper and cardboard for recycling. Consider also how you will get the various types of waste to their final destination. An external door from your kitchen in close proximity to the worm farm/compost bin and the council recycling bin will be a great incentive to minimise what you put in your wheelie bin or rubbish bag each week. If compost and recycling bins are hard to get to from your kitchen, you may find your enthusiasm for using them wanes.

A final thing to remember on the subject of rubbish is that kitchen in-sink waste disposal units are not a good way to get rid of your vegetable scraps and peelings. Aside from using a lot of water, they add large amounts of organic waste to our discarded water, causing nutrient enrichment of our waterways, which in turn results in increased algal growth and reduced water quality.[14]

Bathroom and laundry Although rubbish from the bathroom and laundry is generally given even less consideration than kitchen waste, many of the rubbish solutions that work in the kitchen can also be applied in these areas. Much of the rubbish created in our bathroom(s) and laundry is also recyclable – for example, cardboard laundry powder boxes and empty toilet rolls, recyclable plastic detergent, shampoo and cosmetic containers – and there is no reason why a similar system of sorting bins cannot be included in these areas.

When renovations are planned, the original laundry off the back porch may be sacrificed to make way for a new family room or bedroom – and sometimes rightly so, as utility areas in older homes often occupy the sunniest spot in the house. But the new location of the laundry is likely to have just as much impact on your lifestyle and your carbon footprint as the addition of a sunny new living area, so it is worth thinking carefully before relegating it to a cupboard or basement area a long way from any external doors. If you plan your home so that it is easy and convenient to get from your laundry to your clothesline, you are far more likely to hang your washing out to dry; a laundry on an upstairs floor or in the basement creates a strong temptation to use the clothes dryer (see Case study: Laundry relocation creates a bad habit).

> **CASE STUDY**
> **Laundry relocation creates a bad habit**
> During major renovations to our own home in the late 1990s, we temporarily moved the laundry downstairs into the garage. Previously, I had always hung the washing in a covered area outside to dry, and remember having been amazed when a work colleague told me she always used the clothes dryer because 'I don't have time to be hanging out washing.'
>
> However, after our own laundry was relocated to the garage, popping loads of laundry into the washing machine, and moving them from the washing machine into the dryer, became something I did as I left or returned to the house in my car. It quickly became a major inconvenience to be lugging heavy baskets of wet washing

up the stairs to hang on the washing line – particularly when I knew that, in Auckland's humid climate, heavier items would often need 'finishing off' in the dryer anyway.

Now, when we have clients wanting to locate a laundry in an area that does not have easy access to an outdoor clothesline, we recommend they reconsider – both for the sake of our environment and their electricity bill!

Maintaining healthy homes

Keeping toxic materials out of the environment New Zealand's distance from the rest of the world has resulted in a tradition of using locally sourced timber for framing, cladding and joinery – at least until recently. This makes many New Zealand homes inherently relatively carbon neutral in terms of the life cycle of their materials. The older the home, the more true this is – timber that is structurally sound after 50-100 years or more has met its debt to the environment, and the untreated timber in pre-1970s homes can also be safely reused, recycled or disposed of at the end of the building's life.

The non-toxic construction materials prevalent in older homes also meet other criteria for sustainability, such as providing a quality indoor environment. In fact, the most likely contributors of carbon emissions and toxins in a pre-1950s house are the materials used in later additions and renovations or modern furniture and fittings, which often contain substances such as Polyvinyl Chloride (PVC), Volatile Organic Compounds (VOCs), Halogenated Flame Retardants (HFRs) and Perfluorocarbons (PFCs).[15]

Because many highly toxic and carbon-heavy materials are cheap to produce and buy, and are vigorously marketed to an ill-informed public, it can be a real struggle to convince homeowners to invest in specifying more carbon-neutral and sustainable materials when renovating – even if the cost is only slightly more. This problem is compounded by a large number of tradespeople who also have a poor understanding of sustainability issues, and are reluctant to work with products with which they are unfamiliar. Many builders, for example, will order a house lot of timber treated to the level required for exterior framing because it is easier and more convenient than calculating the separate quantities required for different areas that require less toxic treatment or none at all.

Kitchen and bathroom cabinetry: avoiding the cycle of constant replacement Some time during the last 20–30 years, somebody decided that particleboard, followed closely by its successor, Medium Density Fibreboard (MDF), were good materials from which to make kitchen and bathroom cabinetry. These products were cheap to produce, easy to work with and, best of all, once the melamine or lacquered finishes became chipped or damaged and the MDF substrate got wet (usually within 5–10 years), the MDF would swell and rot – at which point the cabinetry would need replacing. Around the same time, kitchens and bathrooms became 'fashion' items, to be removed and replaced at whim to suit current trends in interior design – and the modern 'cabinet making' industry was born.

In our view, the practice of replacing entire kitchens and bathrooms every 8–10 years (and the use of MDF in these areas) is extremely foolish and wasteful. In addition to the cost and inconvenience involved, it also raises serious environmental and health concerns. Most kitchen and bathroom cabinetry is made out of MDF, which is normally finished with a melamine resin or spray-on lacquer. All of these substances contain formaldehyde and/or other known VOCs, which can be harmful both to workers during construction and people who live with the materials after they are installed, as they release harmful chemicals into the air in a process known as 'off gassing.'[16] Consider, too, what happens to a 'disposable' kitchen when it is dumped in a landfill at the end of its short life. The organic timber component produces methane gas as it decomposes, while the cocktail of chemicals in all of the materials leaches into the surrounding earth, eventually polluting our waterways.

Fortunately, none of these dangers are necessary. Specifying sustainably managed solid timber finished with a natural oil or low VOC water-based product for kitchen and bathroom cabinetry will eliminate all of the above problems, giving you fittings that are safe to live with and dispose of. Solid timber cabinetry generally costs around 15–20 per cent more than formaldehyde-based MDF products, but it will last for 50 years or more, making it much better value for money. This also applies to other furniture and cabinetry you buy or have made for your home.

The biggest problem you are likely to come across with solid timber cabinetry is finding somebody who can make it for you. The rise in popularity of cheap kitset kitchens has coincided with a corresponding decline in the skill of cabinetmakers, and many are now unwilling or unable to work with anything other than MDF and melamine. It will take a massive change in thinking for large numbers of cabinetmakers to return to building in solid timber and other more sustainable materials, and for this reason it is vital that consumers are active and persistent in recreating this demand.

CASE STUDY
The tyranny of the 'almost new' vanity
Our new clients, a couple in their thirties with two young children, invited us to their home to discuss proposed additions and alterations. As we walked around their house together, we were informed that the kitchen and existing bathroom were not part of the brief, as they had already been 'done up' four years earlier, just before their first child was born. We noted that their choice of MDF and melamine cabinetry and benchtops would probably require replacing within another four to six years, and that they may have been better off paying a little more for solid timber cabinetry and stainless steel or granite benchtops.

However, our clients maintained that budget had been a major consideration at the time, and 'at the price we paid, we could afford to replace the kitchen every 5–10 years.' With the subject of the kitchen clearly off limits, we did not persist in discussing the environmental cost of disposing of an entire kitchen every 8–10 years, or mentioning the health hazards of living with new MDF and melamine, both of which emit high levels of formaldehyde and other VOCs.

Several months later, once design work for their extensions was well under way, our clients asked if we thought it would be possible to re-use some of their old 1940s solid timber kitchen cabinetry (which they had retained after 'doing up' the kitchen) to make a vanity unit for their new bathroom.

'There is no way we're buying another of those MDF cabinets,' they told us. 'The one in our existing bathroom is not even five years old and it's completely ruined – the melamine has chipped off; the doors have puffed up so they don't close any more; and where one of the feet has come off it's impossible to fix because the substrate has completely disintegrated.'

Taking an enlightened approach to lighting
Lighting is an area that is often neglected in the design of both new homes and renovations – but it can have a major impact on both the energy efficiency and the quality of the indoor environment in your home. Ironically, it is not light fittings that are the most important factor in sustainable lighting design – it is maximising the level of natural light. While it is not necessary to paint your entire house in pale colours to achieve a good level of lighting, you can enhance both natural and artificial lighting with appropriate interior colours. For example, using pure white on your ceilings will enhance the value of any lighting which shines upward, and it make sense to choose lighter colours on walls in areas where a high level of light is required.

When selecting light fittings, a truly sustainable approach involves carefully thinking about how you will use the different spaces in your home, then selecting the appropriate type of light for each location. This does not necessarily mean getting rid of original light fittings, or having to settle for a modern look either. For example, many lighting styles from the Victorian and Edwardian periods right through to the 1960s and 1970s can be used with energy-efficient bulbs, and there are many original and reproduction examples of lights from these periods available.

Modern lighting can also be used very effectively in renovations. For example, the latest LED lights use only a tiny amount of energy and can be used very creatively for mood lighting and in areas where only a low level of light is required. Because LEDs are so tiny, they can be positioned discreetly under kitchen cabinetry or above pelmets without compromising the integrity of a character home's interior. Adding modern electronic technology such as sensors, timers and dimmers is another way of minimising the energy your home consumes through lighting.

Finally, a few words on recessed halogen downlights. Halogen bulbs are ideal for reading and other close tasks because they create a small, concentrated pool of bright, clear light, and a single bulb is reasonably energy efficient. However, when halogen downlights are used as the main light source for a room, or even a whole house, this becomes much less energy efficient because so many units are required. This style of lighting will also often result in very uneven lux levels – unpleasant glare spots where the light is too bright, and other areas where there is not enough light. Furthermore, recessed ceiling lighting can compromise waterproofing details and/or the integrity of ceiling insulation. When the heat from a halogen bulb rises, it pulls more warm air out of the room through the uninsulated area around the light fitting. In villas, bungalows and Art Deco homes, recessed down lighting is particularly inappropriate, as it spoils the appearance of original decorative ceilings. While it may have its place in work or reading areas where bright pure light is required, recessed halogen lighting is not an efficient or appropriate option for an entire house.

Flexibility for the future
When looking at renovating, many people consider the needs of the current occupants of the home, right now. Often the house is approached as a standalone unit, independent of its neighbours or the local community, and any additions are similarly considered in isolation from the existing house. While this might seem to be the cheapest way of approaching a renovation project, in the long term it can work out a lot more expensive – both for the owners of the home and the environment.

Once again, grand planning – even if the work is to done over several years – can allow you to spend your money wisely and avoid having to undo work or move on when the house once again does not meet your needs. Planning for the future also helps to reduce the carbon footprint of your home both in the construction materials used and discarded, and in the way the home is used during its life.

Room to move You might need to add two extra bedrooms, a second bathroom and a teenage lounge now, but what happens when the teenagers have left home? A well-planned renovation will have created spaces that are flexible enough to be used for several purposes. For example, sometime in the future, the teenage wing could be converted into a self-contained apartment to generate additional income or accommodate extended family members, a home office to reduce travel time and costs, or space for complex craft or hobby projects to be left undisturbed.

Spaces can also function in different ways concurrently. A study or office could double as a guest bedroom, a garage could also be used as a gym or rumpus room, the laundry could incorporate a sewing or gardening room, doors could be installed between two or more living areas to accommodate a large gathering or several smaller ones depending on the occasion.

Perhaps surprisingly, the distribution and location of bathrooms can be a key factor in making your home flexible. While ensuites are considered highly desirable or even essential by many people, providing a bathroom that relates to several rooms will often make the spaces more versatile in both the short and long term. For example, a self-contained studio area of two large rooms and its own kitchen and bathroom would make a more versatile home office space if the bathroom was completely independent of both rooms. If the bathroom is an ensuite to one of the rooms, staff and visitors may have to walk through the boardroom or lunchroom every time they want to use the toilet. Because bathrooms are also very resource-hungry in terms of materials and energy consumption, having fewer of them will reduce the carbon footprint of your home.

Outside, remember not to occupy your entire property with building or paving – leave room for trees, plants and lawn as well. Even producing a small amount of fruit and vegetables on your property helps to reduce your carbon footprint. Placing paths and driveways on the front of the property will reduce the amount of paving required, and selecting pebbles, shells or other permeable materials instead of concrete will allow rain to soak into the ground rather than flooding sewers and drains.

Modern versus timeless While some people believe that additions and alterations should 'make a statement' and stand out from the original building, we have spoken to too many people keen to be rid of renovations that were fashionable 30, 20 or even 10 years ago to support this idea. Making a statement might have seemed exciting at the time, but today, most renovations done in the 1990s, '80s and '70s (and earlier) just look tacky and incongruous – and they are often structurally unsound as well (see Case study: The legacy of the late 1980s lean-tos).

Similarly, removing original features from and 'modernising' quality older homes (including those built in the 1970s and even later) may seem like an easy and convenient alternative to more sympathetic repairs and maintenance, but in the long term doing this is likely to devalue the house. Consider the value today of a bungalow that was modernised in the 1960s or '70s compared with a similar home that still has the original (well maintained) fireplace, doors, windows, ceilings and other features. Unfortunately, modernising is often cheaper and less time consuming than maintaining the original features, and consequently it is a technique favoured by property speculators who want to 'do up' a house for a quick sale. It is, however, a wasteful practice, as it often involves dumping sound materials that still have many years of useful life in them and replacing them with less durable ones that will definitely need replacing in the next 5–10 years.

On the other hand, where a home is renovated using a style of building, materials, joinery and other fittings to match the existing home, the integrity, and therefore the value, of the house is maintained. This approach also provides far greater flexibility for future changes, involves much less (financial and environmental) cost in fixing up previous 'statements' before moving forward, and it helps to retain the integrity of the neighbourhood and local community.

> **CASE STUDY**
> **The legacy of the late 1980s lean-tos**
> Our clients had just purchased a beautiful 100-year-old villa that was listed as a Heritage Building under the District Plan, and they asked for our help to replace the chipped formica and particleboard kitchen and bathroom that had been installed in the 1980s. These were contained in two lean-tos added to the rear of the house, both of which were completely unsympathetic, and were now in a very poor state of repair. By contrast, the original home was unchanged and in good condition.
>
> During the concept design phase, it became apparent that the previous renovations would need to be demolished and completely rebuilt in order to adhere to Heritage guidelines and allow a sensible and functional kitchen layout that met our clients' needs. The additional cost of this would use up almost all of the budget allocated for the entire works.
>
> Had the 1980s renovations been done properly in the first place, our clients' proposed kitchen and bathroom work could have been done easily within their budget (or perhaps would not have needed doing at all). However, because the cost of putting right the poorly planned and constructed later additions was so high, their budget had to be substantially increased. This is in addition to the environmental cost of disposing of the materials used in the lean-to additions and obtaining new materials for building.

Staying put may make it easier to get around One of the key ways in which we impact upon the environment as individuals is through the carbon we generate as a result of motor vehicle travel (see 'Thinking outside the car' chapter). For urban dwellers, this often means everything from getting to work and school or attending sports and social events to the weekly (or more frequent) grocery shopping. So, it makes a lot of sense to design our homes to minimise the need for vehicle travel and/or make alternative means of transport more attractive.

We have come across a number of people who, put off by the apparently high cost of expanding or renovating their home to make it more suitable for their needs, decide to sell and buy elsewhere to achieve the additional space and facilities they require. However, moving to a new dream home further away from established school, work and/or community links can soon become a nightmare with the constant driving to places you could previously walk or cycle to. If you are well established in a community, it may be a far more carbon- (and cost-) conscious decision to spend the same or more on additions and alterations to your existing home than buying another home in a different part of town.

When planning renovations involving garages and/or sheds, try to provide easy and safe access to the street for bikes as well as cars. If you have to trudge across the muddy back lawn and dig your bicycle out of the garden shed every time you want to use it, you are much more likely to take the car. If the bike is in an easily accessible place within the internal access garage, it will be just as quick and easy as driving to cycle down to the dairy or supermarket for a few supplies. With electric scooters now available, and the possibility of larger electric vehicles likely, any new garaging should also include sufficient conveniently located power points to accommodate current and future vehicle recharging needs.

Taking advantage of technology

Technology provides many opportunities for us to make our homes more carbon neutral, but our appetite for more and more electronic gadgets powered by electricity is also adding to the burden our homes place on the environment. So how can we use the 'good' technology to compensate for our reliance on the rest and come out, if not better off, then at least in a neutral position?

Automation technology Simple and inexpensive devices such as timers and thermostats can help us to reduce the energy required to run our homes, as can more sophisticated home automation systems. While home automation technology has the potential to reduce our energy consumption by using sensors and timers to turn off lighting, appliances and even entire zones of your house when they are not in use, there is a high cost associated with this type of technology, both in the equipment and wiring required for its installation, and the computer system that must be run day and night to control it. If it is used just to control lighting and appliances, the running costs are likely to outweigh the energy savings. However, we believe that home automation technology has a lot of untapped potential for managing passive heating and cooling functions which may make it a more economic and carbon-friendly option (see Case study: Looking at home automation in a different light).

A positive way in which technology can help us reduce our carbon footprint is by providing opportunities for us to do more things from home. This includes working, shopping, banking, obtaining information and even socialising. Once again, there is a carbon cost associated with producing and running the computers and other equipment required to do this, but unless all of these facilities are within walking or cycling distance of your home (and you actually do walk or cycle to them), it is likely to be less than the cost of driving there.

Planning is once again an important factor in how successfully technology can be applied in your home, and this is particularly true for renovations. Consideration should be given to the location of wiring and other sources of electromagnetic radiation to avoid areas where people sleep or spend large amounts of time. Automation systems should be considered, but unless the renovation is of a large scale and wall linings are being removed throughout much of the house, it may not be practical or economic to install a useful automated system. In this case, it is better to continue turning off lights and appliances by hand. As manufacturers continue to improve and develop products, it is also likely that sensor technology and automatic shut-off functions will be built in to many devices.

CASE STUDY
Looking at home automation in a different light
We recently visited a supplier of home automation systems to find out how the technology might be applied to naturally ventilating and/or retaining heat in a home or office building. The system we had in mind would measure inside and outside temperatures, weather

and wind conditions, and automatically open and close louvres and/ or drapes to release or retain heat as required to maintain a comfortable temperature.

However, when we suggested that the automation technology had potential as a temperature control device, the salesperson, who had been busily pointing out how the lights came on and off as we walked through the show home, looked blank. After a few minutes' consideration, he suggested, 'Well, I suppose you could get it to turn on the air conditioning if you wanted.'

Solar water heating Possibly the most important technology available to us at present for reducing the energy consumption, and therefore the carbon footprint, of our homes is that which enables us to harness the renewable power of the sun to heat water and run our homes. It is difficult to conceive of an instance where a solar cylinder would not be the best way of providing the bulk of a home's hot water and, in our view, installing a solar powered system should be mandatory for any renovations involving changes to or the addition of a hot water system. While some solar water heating systems include large tanks that may intrusively protrude from the roofline of a bungalow or villa, there are also systems that allow the tank to be concealed inside the roof cavity, with only the solar panels visible on the roof.

Electricity generation Ironically, most of the technology available to help us reduce our energy consumption also uses energy — usually electricity. And as long as our home is consuming even a small amount of energy, can we really claim that it is carbon neutral? Currently, although around two-thirds of New Zealand's electricity production is from renewable sources, the remaining third is not, and if electricity demand continues to rise, we need to consider whether it is still appropriate to depend on large providers to meet this demand, or whether we should be taking responsibility ourselves for at least some of our consumption. Providing small-scale facilities for neighbourhood generation using photovoltaic panels or small wind turbines is more efficient than building large new power stations because there is less energy lost in transit. However, there are a number of barriers preventing the implementation of such a system.

The major barrier currently for consumers who are considering the inclusion of their own power generation systems in a building or renovation project is 'payback'. That is, unless the home is remote and there is no current connection to the grid, the cost of buying and installing the system is more than the expected savings to be made. In our view, this is not necessarily a good reason to eliminate power generation from your plans. Many people install a $20,000 spa bath or buy a giant plasma television with no expectation of recouping any of the expenditure, and with the knowledge that these items will be worthless in just a few years. A domestic electricity generation system, by contrast, will reduce household running costs on an ongoing basis as well as reducing the carbon footprint of the home, and is an equally valid way of spending this amount of money. As electricity prices continue to rise and the cost of photovoltaic and wind technology falls, it will become a more attractive option, even for those who are focused on getting a financial return on their investment. Ideally, our attitudes will also change so that owning consumer goods such as giant plasma TVs will become less of a status symbol than it is today, and there will be more value attached to activities such as operating your own renewable power generation system.

Once we overcome the issues of cost and attitude, we are still limited in the extent to which we can contribute to meeting our own electricity needs by the availability of items such as photovoltaic panels and wind turbines that are small and quiet enough to run in an urban environment. Currently, worldwide production of these items is insufficient to meet demand, and the main producers, including Germany, Japan and the USA are using most of what they produce in their own countries. However, New Zealand is progressing rapidly in these areas. For example:

- Solar cell technology recently developed by Massey University's Nanomaterials Research Centre may soon allow us to produce our own photovoltaic roofing materials, wall panels, glazing or paints – at a tenth of the cost of the silicon-based photovoltaic solar cells available from overseas.[17]
- Leading New Zealand energy infrastructure group Vector is currently undertaking trials of award-winning 'Swift' microturbines with a view to making them commercially available in New Zealand.[18]

The final difficulty with consumers embracing electricity generation on a domestic scale is how this can be successfully aligned with the goals of the power companies who make their profits from controlling the national grid and supply. On the one hand, when individual households and businesses contribute to the national supply, this will reduce pressure on electricity suppliers to create more large-scale generation facilities. In addition, producing electricity in urban areas where it is actually required is more efficient because less energy is lost in transit. On the other hand, if consumers begin producing energy in any significant numbers, this will undoubtedly pose a threat to the traditional relationship between power companies and their clients – one which many power companies currently seem ill prepared to deal with (see Case study: The road to self-powering was disempowering).

CASE STUDY
The road to self-powering was disempowering
When we decided to install an experimental grid-connected photovoltaic power system on our urban property, our solar panel supplier advised us that there was only one electricity provider in New Zealand who would pay us a 'net amount' (the same rate we pay them for receiving electricity) for any surplus electricity we fed back into the national grid. What is more, this provider would allow us to connect to the national grid and receive payment only if we were members of the Sustainable Electricity Association NZ (SEANZ).[19] Even though we had been told this, part of the experiment was to find out how easy it would be to connect to the grid, and what financial benefits might be involved in selling power back to a retailer, so we decided to contact a number of electricity retailers who supplied the central Auckland area and find out what their procedures and purchase rates were.

Surprisingly, on first contact not one of the electricity retailers (including the one endorsed by our supplier and SEANZ) claimed to have any knowledge or information about grid connection at all. It was only after a considerable degree of insistence and perseverance on our part that we achieved the following results.

Our existing provider suddenly produced, when we had finally threatened to switch companies if they were unable to assist, an information pack that included forms to complete and the amounts they would pay for our surplus generation. This had obviously been professionally prepared and available all along. Unfortunately, the payment offered was so low (around 20 per cent of what we were paying them) that we decided to switch to another company anyway.

The electricity retailer recommended by our supplier and SEANZ eventually gave us the contact details of the person within the organisation who dealt with grid connection and the retailer did indeed agree to pay us a net price per kilowatt for whatever we fed back onto the grid. While it was reasonably difficult to make contact with the right person and obtain the required paperwork to complete, once this was done the actual grid connection process was simple, and the system has run seamlessly ever since.

Other power companies with whom we had no existing relationship remained generally unresponsive to our repeated requests, and continued to advise that they did not have facilities or systems in place to allow them to accept or buy back power off their customers. Perhaps if we had been existing customers that were threatening to move to another provider, these companies too may have suddenly been able to provide more information.

Changing how we think

So far, we have discussed a number of practical ways to reduce your home's carbon footprint as part of a renovation. Most of these are easily achievable right now, and will not add a lot (if anything) to the cost of a renovation. So why are we not doing these things? We believe a lot of simply comes down to habit – and it is time to change.

New Zealand's tradition of high levels of home ownership and low levels of personal savings, coupled with our tendency to move house frequently seems to have resulted in many people viewing the house they live in as a money-making investment, rather than a home. In addition to the home they live in, many New Zealanders also own houses purely as investments – on these they spend the absolute minimum to make them look presentable, leaving inherent problems such as poor heating and ventilation, high energy use, inferior quality materials or inflexible design for tenants or the next owners to deal with.

When buying homes, we are often equally guilty – happy to pay a premium for a new kitchen and bathroom that might last 5–10 years, but assigning little or no value to durable non-toxic materials, quality design that provides passive heating and ventilation or flexible spaces that will accommodate our needs many decades into the future. This problem is often compounded by those in the business of selling property who, in response to the perceived market, will advise clients to renovate a kitchen or bathroom or add 'whiz-bang' technology to improve a home's saleability (see Case study: Perceived value). These 'improvements' are often done using cheap, poor-quality materials that have a short life expectancy and contain high levels of PVC, VOCs and other toxins.

> **CASE STUDY**
> **Perceived value**
> We recently consulted a real estate agent and a valuer on behalf of one of our clients. This client was planning major renovations to his home in an upmarket Auckland suburb, and wanted some idea of what his house might be worth on completion of the work. Both the valuer and the real estate agent expressed the opinion that potential buyers would see little value in planned features such as passive and active solar heating, rainwater collection, non-toxic and environmentally friendly materials and finishing. Instead, they suggested our client would be better off incorporating a swimming pool, sophisticated home theatre system, automated lighting and electronics systems, and additional garaging.

TACKLING THE CHALLENGES

Now that we understand the challenges in renovating to a carbon neutral standard, how do we meet those challenges? The good news is, we have already started along the path, and some of the initiatives currently in place provide exactly the right foundation to help us achieve our goals in the near future.

Current initiatives

While there has recently been a groundswell of awareness about sustainable building, much of the talk has not yet translated into action. Of the action that has been taken so far in New Zealand, much has related to commercial and civic buildings. But there have recently been some promising resources, tools and regulations developed which can also be applied to making our private residences more sustainable – some of these are outlined below.

Resources Smarter Homes[20]: This website was launched in July 2007 and provides clear, independent, factual information about sustainable home design, building and lifestyle options. Aimed at home owners and renters, and at building and property professionals, Smarter Homes was created by a team including the Department of Building and Housing, the Ministry for the Environment, the Consumers' Institute, Beacon Pathway Ltd and URS.

Level[21]: Also launched in July 2007, this website is designed to assist architects, designers, builders and others involved in the construction industry to design and build homes which have less impact on the environment and are healthier, more comfortable, and have lower running costs. This site is developed and maintained by the Building Research Institute of New Zealand Ltd (BRANZ).

GreenBuild[22]: Launched in October 2007, this website aims to provide independent technical and environmental vetting data for a large number of individual products available to the New Zealand design-construction industry.

Tools Green Star NZ[23]: This suite of rating tools is being developed by the New Zealand Green Building Council. The tools will provide 'New Zealand's first comprehensive environmental rating system for buildings.' While the only version currently available (released in April 2007) is for office buildings, the Green Star NZ rating tool system will eventually be developed for different building types such as retail, health, education, residential, industrial and so on, and work is already in progress on a number of these.

HERS[24]: A Home Energy Rating Scheme (HERS) is currently being developed by the Energy Efficiency and Conservation Authority (EECA) in consultation with the Department of Building and Housing and the Ministry for the Environment. The scheme will assess a home's energy performance over a range of criteria, with the resulting report providing home owners with information about how they can improve their home's energy efficiency, and prospective buyers with a valuable tool to help them assess the ongoing costs of running the home. Implementation for the HERS is currently scheduled for December 2007 on a voluntary basis, although it is likely to become mandatory at some stage for anybody who is selling or renting out a house/flat.

TUSC[25] (Tools for Urban Sustainability: Code of Practice): This is a web-based analysis tool to plan new urban developments and assess them against sustainability indices. The tool is being developed as part of a two-year project initiated with the assistance of Waitakere City Council and the Ministry for the Environment Sustainable Management Fund.

Regulations Building Act 2004: A key purpose of the 2004 revision of the Building Act is that 'buildings are designed, constructed and used in ways that promote sustainable development. The NZ Building Code [now] requires that designers, builders, local authorities and building owners consider:

- Minimising waste during construction
- Use of sustainable materials
- Use of safe and healthy materials
- Energy conservation and efficiency
- Material durability (designers must make homeowners aware of the maintenance requirements of the materials specified and that whole of life costs are a better determining factor in material selection rather than just initial cost).'[26]

Insulation: Minimum standards of ceiling and wall insulation have been mandatory in all new buildings and renovations since 1978. New rules requiring higher minimum R values of insulation and the use of double glazing for many areas are (at the time of writing) being developed, with decisions and timeframes for implementing these new regulations being announced from October 2007.[27]

Solar Water Heating: A government subsidy of up to $500 is available towards the cost of the installation of selected solar water heating systems. The government is also assisting with industry and public education campaigns to promote the use of solar water heating, and providing guidelines for local authorities to help reduce the cost of installing solar water heating systems.[28]

Future vision

Over time, the regulations and tools discussed above will probably succeed in bringing about the changes we need in order to create carbon-neutral buildings. Change is starting to happen now, slowly – but time is a luxury we do not have. So how can we speed up change in order to achieve our goal of carbon neutrality by 2020? We believe the responsibility lies with three key groups – consumers; building and property professionals; and local and national authorities.

Consumers Out of the three groups, consumers may find it easiest to sit back and wait for the other groups to take action, letting government, local bodies and building professionals tell us what to do to make our homes carbon neutral. However, in many ways we also have the greatest power over how, and how quickly, change happens.

The first thing we must do as consumers is to be informed – or work with someone who is. Reading books like this one, and accessing websites like SmarterHomes is a good start in either case.

If you are capable and prepared to 'do it yourself,' you should also research thoroughly any products and materials you are considering using in your home – where they come from, their embodied energy, what harmful chemicals they may contain, how they may influence the indoor environment during use and/or harm the wider environment after their disposal. If a product does not stack up, do not buy it – no matter how temptingly low the purchase price is.

If you work with architects or designers, you should be sufficiently informed to evaluate their capability (or lack or it) to assist you with developing a sustainable home, and to work with them in a meaningful way. 'Sustainability' is a buzzword in the design and construction industries at present, and while there are many people who talk about it, there are far fewer whose knowledge is more than superficial, and who actually apply sustainable principles to their design work. When you do find the right person, you should also be prepared to implement their recommendations – and accept that their fees and the up-front cost of their designs and specified materials will not necessarily be the cheapest option available. The gains will be long term, and they may be environmental as well as (or even instead of) financial.

Secondly, whether renting or buying, we as consumers must change what we value about our homes, and be prepared to pay (or not) accordingly. Table 1 shows a number of items that might be included in a real estate flyer advertising a home for rent or sale. The items in the left column are currently popular selling points, although there is a high financial and/or environmental cost involved in either running or regularly replacing them. The items in the right column can be obtained for a similar cost to their counterparts in the first. They will last a long time, reduce the home's ecological footprint and save you money over the life of the house, yet they are not features that are regularly promoted to appeal to prospective home buyers or renters.

WHAT WE CURRENTLY VALUE	WHAT WE COULD HAVE FOR A SIMILAR COST
Modernised throughout with aluminium joinery, new (MDF) kitchen, 4-car garage, ensuites in three bedrooms	Original character and joinery, solid timber kitchen, 2-car garage + large bike shed, two quality bathrooms, self-contained area
Gas-fired central heating system	Photovoltaic solar panels on the roof
Mains pressure, instantaneous, gas hot water system; spa bath	Solar hot water cylinder
Kitchen in-sink waste disposal unit	Three-bin rubbish-sorting station and worm farm
Polished floors and interior repainted throughout	Ceiling, wall and under-floor insulation at or above the minimum requirement
Spa pool in the back yard	Rainwater-collection system fed to garden taps, toilets and laundry

Table 1 What should we value in a home?

As consumers, by asking to look at homes that include items in the right column, we can and will influence what is considered valuable and important – both by real estate agents, developers and others who buy and sell homes.

Thirdly, we must abandon our fear of 'over capitalisation' and see investing in our homes as a way to enhance the health and happiness of families, even generations, over a longer period of time. If we can overcome our culture of frequent moving, we will be far less likely to 'do up' a home cheaply in order to sell it for a profit – and also be less likely to purchase homes that are not efficient or sustainable if we do need to move.

Building and property professionals Like consumers, building and property professionals have a responsibility to be informed – but even more so. In addition to understanding the basics about designing, building and living sustainably, we must also: have a thorough knowledge of products, materials, methods and design features that make a home more (or less) efficient and sustainable; be able to explain this to others who ask our advice; and be able to use this knowledge to design, build and promote carbon-neutral sustainable homes. People in different professions will focus more strongly on different areas. For example:

Architects and designers will be very familiar with the properties of various building materials and will specify appropriate ones. They will understand concepts such as thermal massing and passive ventilation and include these in their designs as a matter of course. They will embrace new technologies such as Building Information Modelling (BIM), which will help them to better plan appropriate spaces, analyse sun angles and minimise wastage when construction begins.

Builders and other tradespeople will use information provided from BIM to calculate and order materials in a way that minimises wastage. They will reuse demolition materials on site where possible, and sort all other waste for appropriate recycling and disposal. They will engage with architects, designers and their clients to learn about new, more environmentally friendly products and use them successfully.

Plumbers and electricians will understand how solar water heating, photovoltaic electricity and rainwater-collection systems work, and know how to install them efficiently and correctly. They will also be familiar with the pros and cons of the different materials they work with, and know to avoid toxic materials such as PVC or advise clients on issues such as selecting energy-efficient lighting or water conserving tapware.

Product suppliers and manufacturers will evaluate both new and existing products against environmentally sustainable criteria and delete items and ranges that have high embodied or ongoing energy costs or are not sustainably produced. They will provide comprehensive and detailed information to architects, designers and tradespeople to allow them to make decisions about products and ensure that their products are installed and used correctly. They will advertise their products in a way that encourages consumers to consider issues such as energy and water consumption and product life cycle, rather than focusing on up-front cost.

Real estate agents will draw attention to features such as low energy consumption and eco-friendly water collection or waste disposal systems when marketing homes. They will discourage cheap and quick fixes that adversely affect the environmental footprint of a home when advising clients on how to improve the saleability of the property. They will encourage potential buyers to evaluate the energy consumption and materials of a home as a higher priority than things like the number of bathrooms and whether a home has an alarm system.

Training providers will support these professionals by ensuring that new graduates are aware of and understand the importance of building and developing sustainably, and know where to find the information they need to keep abreast of new developments. Professional organisations will help their members remain up-to-date by supporting initiatives such as the GreenBuild website and providing relevant seminars and workshops as part of continuing professional development programmes.

Local and national authorities It would be nice to think that consumers and building professionals will inform themselves and voluntarily implement the building and renovation practices that will help to make us carbon neutral by 2020. However, the reality is that more formal systems and rules will also need to be in place to drive this change. In order to ensure that the information available is used effectively, our view is that some or all of the following will need to be in place at a national and/or local government level.

The information on websites such as SmarterHomes and Level will be available to a wide audience. The websites will be promoted via local authorities as part of the building consent application enquiry process, and they will be advertised widely on television, radio and in print as well as via the Internet. While the electronic version is the ideal medium for both sustainability and currency, printed options should also be available to those who are not especially computer literate or have no Internet access. Publication of the website content in serial form – for example, in lifestyle magazines, industry publications and/or newspaper supplements – will allow it to be disseminated in small, manageable chunks.

Consumer labelling and product details, perhaps similar to the Heart Foundation's 'Healthy Choice' labelling system will be widely available to assist both professionals and lay people choose sustainable building materials and fittings. 'Environmental Choice' is the official government scheme,[29] and a 'Green Tick' system has also been introduced by a private organisation,[30] but there has been little uptake for either system with building materials to date. The new GreenBuild website may succeed where the others appear to have not – if it does not, more rigorous and formal means may be required to encourage consumers and building professionals to embrace a more sustainable approach to building and renovation.

Real estate agents and anyone selling or renting out a house will be required to provide an independent report to prospective purchasers or tenants, allowing them to assess features such as the home's energy performance, life expectancy of materials and indoor air quality. The Green Star and HERS systems both address this to a certain degree, but it would be preferable to have a single, commonly accepted evaluation system that addresses more than just energy performance. This will ensure that developers and others who build and renovate for profit use healthy and sustainable materials that will score well in an assessment – failure to do so will lower the value of their investment.

Accessible facilities will be available to support the recycling and reuse of building site waste. The dumping costs of building and other waste will be substantially higher than they are now, which will make it uneconomic to dump rubbish such as glass, metals, timber and concrete that could be reused or recycled (see Case study: The waste bin recyclers who were not).

CASE STUDY
The waste bin recyclers who were not
Preparing for a small renovation on our own property, we agreed with the builders that we would arrange a bin for any waste materials. After contacting a large number of jumbo bin and rubbish skip providers, we finally found a very reasonably priced company who confirmed that they had a recycling policy which included sorting all rubbish they collected into various categories of materials for recycling, thereby reducing the amount of rubbish being taken to a landfill.

On completion of the work, we commented to the driver collecting the bin that we had had difficulty finding a company committed to recycling. 'Oh, yeah,' he told us. 'We used to sort everything at the yard, and it was a great system – saved a lot of money too – that's why we have always been cheaper than the others. But we've just been bought out and the new owners have scrapped that system. They're just putting up the prices a bit to cover tip fees and throwing the lot on the dump.'

The problem here lies not just in the attitude of the new company owners, but also in the rules and fees set by the local authority. If the dumping fees were substantially higher (rather than just marginally more) than the cost of recycling, there would be a greater financial incentive for rubbish removal firms to recycle rather than dump construction waste as the former would become a more profitable option. The local authority could also just refuse to accept recyclable materials for dumping, forcing rubbish removal companies to recycle where possible.

The sustainable development clauses of the new Building Act 2004 will be consistently and rigorously implemented by local authorities. Perhaps similar to the risk matrix for weather tightness, a 'sustainability matrix' would evaluate elements of a building's orientation, design and materials, projected water and energy use to come up with an overall score that would either be accepted or rejected by the local council. The Green Star or TUSC tools may be useful starting points.

The Building Act will be further amended to discourage the use of heavily treated, toxic materials in new homes and other buildings, rather than requiring the use of these materials (especially timber) as it does currently (see Case study: Turning timber from the obvious choice to a noxious choice).

> **CASE STUDY**
> **Turning timber from the obvious choice to a noxious choice**
> In response to problems with leaky homes, New Zealand building standards were amended to include higher timber-treatment levels, and the use of treated timbers in areas where previously no treatment was required. While this may help to avoid or delay the incidence of rotting timbers in homes with particular design features and/or claddings, its overall effect has been to require more chemicals to be introduced into all of our timber-framed buildings. In addition, dealing with the by-products of timber-treatment processes and eventually disposing of the treated timbers will pose wider and ongoing environmental challenges.
>
> A more appropriate solution might have been to increase timber-treatment levels to cover only buildings with a high risk of leaking because of their design and/or claddings. This would have the dual advantage of allowing the many buildings that do not have high risk factors to be made from more environmentally friendly materials, while also discouraging designs and materials that are prone to problems as a result of leaking.

Import controls and/or tariffs will be introduced to ensure that only sustainable building products are imported to and marketed in New Zealand. Many timbers, joinery, claddings, flooring, cabinetry and other building materials are imported from other countries, and although they usually must meet safety and durability standards, there appear to be no specific regulations governing the sustainability of these products.

THE 'PRESTIGE' HOME OF THE FUTURE

Once we all have the knowledge and the resources to build sustainably, and using them becomes ingrained in the Kiwi way of doing things, we will have arrived at our new paradigm in home renovation – and a new definition of prestige for our homes. To conclude this chapter, let us consider just a few examples of what this definition might include.

- The design of your home allows you to maintain a healthy indoor temperature all year round, with no additional energy needed for heating or cooling.
- The energy requirements of all appliances, lighting and other electronic equipment used in your home are met from renewable sources on your own property.
- Your home is made entirely from materials grown and/or manufactured in New Zealand.
- All the materials used in your home are made from renewable resources, and they will last as long as or longer than it will take to grow or produce those resources again.
- Everything in your home can be reused or recycled, or will decompose safely at the end of its useful life.
- Every room in your home can be used for more than one purpose (well, maybe not 'the little room'!).

ENDNOTES
1. Wiedmann, T. and Minx, J. 2001. *A Definition of 'Carbon Footprint'*. ISA UK Research & Consulting. http://www.isa-research.co.uk/docs/ISA-UK_Report_07-01_carbon_footprint.pdf (accessed 29 Aug 07).
2. http://www.mfe.govt.nz/issues/sustainable-industry/govt3/topic-areas/sustainable-buildings/what-is.html (accessed 29 Aug 07).
3. http://www.mfe.govt.nz/issues/sustainable-industry/govt3/topic-areas/sustainable-buildings/design.html#materials (accessed 29 Aug 07).
4. The rating system used to indicate the ability of the insulation material to stop heat transfer.
5. http://www.ecoinsulation.co.nz/wa.asp?idWebPage=4080 (accessed 29 Aug 07).

6   http://www.terralana.co.nz/faqs/ (accessed 29 Aug 07).
7   http://www.pilkington.co.nz/applications/products/asia+and+australasia/new+zealand/english/bybenefit/solar+control/products/pilkington+comfortplus/benefits.htm (accessed 29 Aug 07).
8   http://resene.co.nz/pdf/Sustainability.pdf (accessed 29 Aug 07).
9   http://www.energystar.govt.nz/index.aspx (accessed 29 Aug 07).
10 http://www.energyrating.gov.au/index.html (accessed 29 Aug 07).
11 http://search.waterrating.com.au/ (accessed 29 Aug 07).
12 Mobbs, M. 1998. *Sustainable House*. Sydney: Australian Consumers' Association, CHOICE Books.
13 http://www.waitakere.govt.nz/abtcit/ec/bldsus/pdf/water/savingwtr.pdf (accessed 29 Aug 07).
14 http://www.niwascience.co.nz/ncwr/wru/2005-13/logging (accessed 29 Aug 07).
15 http://www.healthybuilding.net/pdf/Healthy_Building_Material_Resources.pdf (accessed 29 Aug 07).
16 http://www.wisegeek.com/what-are-the-health-risks-of-mdf.htm (accessed 29 Aug 07).
17 http://www.renewableenergyaccess.com/rea/news/story?id=48187 (accessed 29 Aug 07).
18 http://www.vector.co.nz/news/208 (accessed 29 Aug 07).
19 http://www.photovoltaics.org.nz/index.html (accessed 29 Aug 07).
20 www.smarterhomes.org.nz (accessed 29 Aug 07).
21 www.level.org.nz (accessed 29 Aug 07).
22 www.greenbuild.co.nz (accessed 29 Aug 07).
23 http://www.nzgbc.org.nz/index.php?option=com_content&task=blogcategory&id=80&Itemid=75 (accessed 29 Aug 07).
24 http://www.eeca.govt.nz/residential/home-energy-rating-scheme/indexnew.html (accessed 29 Aug 07).
25 http://tusc.synergine.com (accessed 29 Aug 07).
26 http://www.level.org.nz/material-use (accessed 29 Aug 07).
27 http://www.beehive.govt.nz/ViewDocument.aspx?DocumentID=29180 (accessed 29 Aug 07).
28 http://www.eeca.govt.nz/news/media-releases/500-grants-for-swh-launched.html (accessed 29 Aug 07).
29 http://www.enviro-choice.org.nz/index.html (accessed 29 Aug 07).
30 http://www.greentick.com/index.html (accessed 29 Aug 07).

Carbon neutral living
in the typical New Zealand house

Brenda and Robert Vale

The authors are Professorial Research Fellows at Victoria University, Wellington. They have been working on issues related to the sustainability of the built environment for over 30 years. They have received a number of awards for low energy and solar buildings, including the UK's first autonomous house and the first zero-emissions community. Both are members of the United Nations Environment Programme Global 500 Forum for environmental achievement.

THE CURRENT SITUATION
How we live in our houses has a major impact on our carbon footprint and hence on the environment. In the past, in New Zealand the home was seen as a place of production – of food, clothing, toys and even furnishings. Now houses are places where growing numbers of Kiwis practice overconsumption, often conspicuously, with a consequently large carbon footprint. Goods, many unnecessary and purchased without thought for the environment, fill homes; and behaviours and activities around the home use resources despite the existence of carbon-neutral alternatives. One way forward is to change the decisions that are made at home, making the home more productive and less a place of overconsumption and waste.

THE VISION FOR 2020
A sustainable New Zealand will be one where the items, activities and behaviours that cost the least will also have the least impact on the environment, and so will be the first choices for most householders. The future will be one of home-based and local production of basic needs such as food and energy, although ideas will be shared internationally. The future will be one where the emphasis is on greater quality in our lives rather than greater quantity.

STRATEGIES TO ACHIEVE THE VISION
Producing our own, reducing unnecessary consumption, sharing and recycling of goods will need to become part of the average New Zealander's way of life. Achieving sustainability will require backing from the government in terms of policies to set New Zealand on the right road, and incentives (like support for insulation and solar energy use, lower road taxes for owners of small cars, encouragement of gardening) to help us all make the right sort of decisions.

INTRODUCTION
Over the years a lot of human energy has gone into considering the technology and techniques to construct a low-energy, low-impact, sustainable house for New Zealand, but less thought is being given to the equally important issue of the environmental impact of how we live in such a house.

Much can be done at home, as this chapter will try and reveal. You do not have to be rich enough to buy carbon credits to reduce your carbon footprint. In fact changing behaviour should be the first thing everyone thinks about and buying carbon credits should be very much a last resort. In the Middle Ages, it was the wealthy sinner who could buy him- or herself out of the problem of eternal damnation, by purchasing from the church what was called an indulgence. In the past, all the poor could do to get into heaven was to behave well. Parallels can be drawn with the current climate change situation, where it is the poor of the world who have to behave well in terms of their impact on the environment, while the rich countries, on the whole, do what they like in terms of emissions and resource grabbing. It is these very same countries who have proposed the idea of carbon credits. Buying their way out of a problem avoids having to change behaviour, whereas it could be argued that what we do, rather than how we atone for it, is the essence of ameliorating climate change.

HOW MUCH CARBON?

People often worry about making a house from something 'sustainable', like rammed earth or straw bales, but in fact the material we use to make our homes is the least important aspect in terms of carbon impacts. Table 1 shows the relative carbon emissions associated with different aspects of our homes and lives as tonnes of CO_2 per year for a four-person household. The figures (based on work carried out at Manaaki Whenua Landcare Research) are approximations, and are intended as a guide only, but what the figures show is that the 'normal' house in New Zealand is not sustainable and that we can do better.

ACTIVITY	CURRENT	POSSIBLE
House construction and maintenance	0.2 tonnes	0.3 tonnes
Waste	1.0 tonnes	0.3 tonnes
House operation (power, light, heating, hot water)	2.0 tonnes	0.0 tonnes
Driving the family car	4.0 tonnes	1.0 tonne
Food (for four people)	16.0 tonnes	4.0 tonnes
One return flight for the family to Europe	16.0 tonnes	0.0 tonnes

Table 1 New Zealand household carbon emissions from different aspects of life in tonnes per year

Note that emissions from house construction and maintenance are higher (by only 0.1 tonne) in the 'possible' column because of greater insulation, double glazing, solar panels, etc. but operating emissions fall by 2.0 tonnes to zero because of the improvements to the house.

The house of the future

What would this mean if we could look into the future, to a time when New Zealand was a truly sustainable society? The homes might not be so different from the timber frame-and-weatherboard cladding houses of today but they would have been fitted with a great deal of insulation in the floor, roof and walls and the windows would be multiple panes of glass with insulated shutters for night time use. Energy for the house would come from solar panels, from small wind generators in the garden and from wood. More food would be grown at home, with food scraps composted. Less consumption would also mean less packaging and less waste generation. The norm would be to walk or cycle to work and school. Long distance travel would be by improved public transport or using an efficient hybrid car. The family's diet would be organic and largely vegetarian. Overseas holidays involving air travel would become an unusual event in anybody's life. The family's annual carbon dioxide emissions would be a mere 5.3 tonnes, compared to the 39.2 tonnes for a family today.

What some of these things would mean in terms of family living is outlined in more detail in the rest of this chapter.

Home is where the heart is

For most people, the house fulfils the basic human need for shelter. However, a typical New Zealand house now does very much more than this. A house, whether rented or purchased (generally with the help of money borrowed from a third party, like a bank), is often said to express an occupier's personality and status. A house can also express the owner's sense of style and fashion, or of course their lack of these. For all of these reasons, houses are often larger and contain many more possessions than might be considered sufficient to provide shelter and support life.

Life could be like a holiday

Think of the difference between the accommodation and associated 'stuff' we use when camping or on holiday in a caravan, and the space and possessions found in the average three-bedroom detached house. There can be lessons learnt from how we live for short periods, usually with good weather, at our favourite destinations. On holiday, the tent or caravan lifestyle is perfectly adequate, and good use is made of the space immediately around the dwelling for activities, providing the much vaunted indoor-outdoor flow beloved of Kiwi designers. Many families enjoy confined holidays of this type, not just those on lower incomes. After all, the space available on a yacht is no greater than the space in many caravans. Holidays in small spaces can force a family to come together and gain entertainment from each other. If ordinary life were much more like life on holiday, the impact on the environment would be reduced considerably in terms of the resources used for shelter and possessions. Maybe the first step for trying to live in a more sustainable way is to make sure a home is no larger than required for the people who live in it – but this is not the whole story.

How we live
For most New Zealanders, this means acknowledging that a household is a dynamic unit that changes over time. In the past, when people lived in extended families, there was a more or less stable need for space. As the children grew up, usually the eldest son and his family would take over the running of the household as the parents became aged and infirm, but all would still live under the same roof. The same would happen with their own children in due course. The multi-generational family home has pretty much disappeared in New Zealand, although there may be examples of it in our Asian population (because of the cultural values attached to looking after and respecting parents) and the tangata whenua (because Maori traditions of living with whanau are still strong in some places).

In general, though, this change in cultural attitudes means that a house with three or more bedrooms may now be lived in by one elderly person, without the support and company offered by the multi-generational family. Let us reconsider the value of the multi-generational family living under one roof. Not only would such a family make maximum use of the resources that go into providing shelter over time, but there would also be sharing of certain resources, such as washing machines and other appliances, rather than separate households (however small) having a full set of modern appliances.

Co-housing schemes are based on these principles. Individual families are allocated a small private house, and central facilities, such as a common dining and guest house, and common laundry and gardens, are supplied. The co-housing movement originated in Denmark 35 years ago and co-housing became popular in both Europe and the USA in urban, suburban and rural settings before coming to New Zealand. An important example here is Earthsong at Ranui in Auckland.[1] This is a community of 32 houses and units, designed using passive solar heating principles, with shared spaces, such as the common house and landscaped gardens where food is grown based on permaculture principles. Meals are cooked and shared in the common house that also has a children's playroom, a sitting room and space for activities such as yoga classes for the adults. What is impressive about the community is the lack of fences, an expression of the sharing ethic, the whole of the outside space becoming a wonderful shared playground for the children.

Even without purpose-built co-housing schemes, it is possible, and might even be inspirational and fun, to have more sharing of resources between neighbouring families who wish to live more sustainably. When the shared washing machine breaks down having two or more heads to gather round and offer suggestions about possible ways to fix it could be so much better than struggling on one's own. As well as reducing environmental impact through the reduction in the number of washing machines made, shipped and run, another benefit of such an approach would be bringing people together and loss of isolation, with the concomitant positive health benefits, and a reduced national health bill.

The importance of maintenance and re-use

Another important behavioural aspect of living sustainably is to adapt to living in an existing house rather than building a new one. Obviously, the longer the resources that have gone into making a house last and continue to provide satisfactory shelter, the better it is for the environment. In Europe, it is not unknown for 500-year-old (or even older) houses still to be giving good service. The key to longevity is good maintenance. Studies by the Building Research Association of New Zealand (BRANZ) have shown that there is a very large backlog of home maintenance in New Zealand. This also puts an onus on designers and manufacturers to produce houses that are both durable and easy to maintain with the skills that are lodged in the general population. New Zealand's tradition of single-storey dwellings makes such maintenance a less hazardous task as most items like gutters and windows (cleaning), and walls (repainting) are within easy reach.

The recent leaky homes scandal showed the waste of resources that results from not designing buildings to last, through omitting details like flashings, drips and sills that throw water clear of the building to prevent it from getting into the structure. Fashion dictated the smooth plastered surfaces that have since been found to survive the climate of New Zealand only if great care is taken in their construction.

Obviously, at some point in the life of the house there will be the need to repair and replace components that are faulty or worn out, meaning new resources will be consumed. This is just part of good maintenance. What is not part of good maintenance is throwing out perfectly good resources in the name of fashion. The number of new kitchens on the market suggests that most people who move into a 'new' house immediately rip out the existing kitchen, discard it, and put in a new one. What is needed is a change of attitude so that a kitchen that is still functional, complete with its drawers and cupboards, is repaired and retained. If it is unattractive, perhaps a superficial renovation such as a new coat of environmentally friendly paint,[2] or some new, possibly recycled, handles for the doors, could achieve the desired look. Not only would this approach be better for the environment, but it would also be much cheaper. And, it would create a unique kitchen, expressive of you and your concern for a sustainable future for New Zealand.

RUNNING THE HOUSE

Apart from the issues of establishing a household, how the house is run on a day-to-day basis will also have an impact.

Turning things off

A small house not only requires fewer resources to build, but will often also use less energy than a larger house not just because it is smaller to heat, but because it has fewer electrical appliances and light fittings. The larger the floor area, the more lights there will be which can be left switched on. We need to change our behaviour, to remember to turn things off when they are no longer needed or wanted. A generation ago, 'turning off' was the only option but with the advent of the remote control, we became able to leave appliances on stand-by, still consuming energy even when not in use. Not only does switching appliances off at the plug and switching lights off when you leave the room reduce our environmental impact, it will also save you money. In a user-pays market economy it is surprising that energy conservation is not uppermost in everyone's thoughts.

Stopping the rot
Because New Zealand is a long thin country in the middle of hundreds of kilometres of sea, and most settlements are on the coast, there is a lot of moisture in our air. Yet dampness created inside a house is just as good a way to rot the materials of a building as moisture and rain coming in from the outside. Living in a house produces moisture from activities such as taking a shower, washing up, cooking and even breathing. If house interiors are warm, the moisture stays in the air and is relatively harmless, but the problem arises because nearly all houses in New Zealand are under-heated. When moisture-laden cool air comes into contact with cold surfaces such as walls and windows, the air is cooled and can no longer hold as much moisture, so condensation forms. The wet surface can lead to mould growth and deterioration of the building's fabric. Mould growth can also lead to health problems. Ensuring that the house is adequately heated throughout, including its surfaces, will help to combat the problem of condensation. And having sufficient insulation of the fabric will mean that the energy required for heating is minimised.

Behaviour also plays a part in moisture control. Drying laundry indoors can add a very high moisture load to the house. Laundry is much better dried outside on the clothesline, or in wet weather on verandas, in carports or in unheated rooms like conservatories. These can be opened up to the outside for ventilation while still sheltering the items from the rain. When wet activities (meaning ones that generate lots of moisture, like boiling pans on the stove, taking a shower or mopping the floor) occur in the house then it is a good idea to ventilate the room to remove the moisture, either by turning on fans in bathrooms or the rangehood in the kitchen, or by opening a window. This is also the reason for sleeping with the bedroom window open at night and ventilating the bedroom and bedding in the morning to remove the moisture that has built up over night. These are all actions that would have been very familiar to your grandmother but that have been forgotten in a world where we prefer to use a machine – the dehumidifier – to remove moisture, most of which we could have avoided putting into the house in the first place.

Water and heating
What the humidifier example shows is that many of us have forgotten how to behave to optimise a house's longevity. A machine has replaced wisdom. Another example is when some renewable and alternative technologies are introduced into houses. A solar water heating system is often plumbed in with an electric immersion heater or gas-fired back-up heater so that the water is always hot, whatever the weather. At first sight, this seems an ideal situation. But a better solution to having hot water when needed may be to have a manually activated electric or gas water booster and a wall-mounted thermometer. When the sky is overcast, if the temperature is too low for comfort then the booster can be manually turned on. In this way conventional energy is used only when it is needed.

The many people who live on tank water (rainwater collected off the roof) in New Zealand, about 10 per cent of the population, are well aware of cause and effect in terms of resource availability. During periods of drought, they automatically conserve water through changes in behaviour. Thus, rather than having the daily shower, which might consume 30 litres, the tank-water user chooses to have an all-over body wash in the basin, which might use 5 litres of water. As water is more likely to be in short supply in the warmer weather of the summer, this is not even going to be a very uncomfortable thing to do. It is a part of living life more as if you are on a camping holiday. This attitude to the use of renewable (solar, wind, rainwater) systems is necessary for the owners of houses fitted with these systems. The user is directly in charge of resources, not distanced from them by being billed for services long after they have been used.

THE IMPACT OF POSSESSIONS
As mentioned above, today's houses are not only larger but are also filled with many more possessions than houses in the past. As little as two generations ago people had, and expected to have, fewer appliances, fewer clothes and a smaller impact on the environment in terms of the resources consumed to support living.

Le Corbusier's garden shed

Modernism positively argued for empty space in the home and much less clutter, to the extent that the architect Le Corbusier, the champion of the modern house and the way of living in it, suggested in his 1923 book *Vers une Architecture* that most possessions should be stored in cupboards with only selected items being brought out to adorn the space or be admired at any one time. The familiar consequence of this is cupboards and garden sheds full of objects that have been put away as being possibly useful but that never again see the light of day.

This behaviour has a double impact, as not only are resources tied up in things that are never used, but resources are also tied up in storing these objects for an indefinite period of time and an unspecified purpose. Sustainable living must, therefore, be linked to thinking carefully about consumption in the home. When, after consideration, an object is purchased, the best thing to do is to look after it so that it has a long and useful life. Because the values of modern society are so geared to the ideas of consumption and change, and because shopping is undeniably fun, some people attempt to satisfy these desires by swapping possessions with others. The popularity of car boot sales, and their electronic equivalents, such as eBay and TradeMe, testify to our desire to satisfy the craving for possessions while having less environmental impact: all we are purchasing are the unwanted goods of other people. Moving resources around in this way is much better for the environment than buying new, providing the habit does not lead to hoarding and the need for even bigger houses.

Avoiding waste

Compared with buying unnecessary things, opting for no packaging comes a poor second in the level of environmental impact. However, reducing packaging is important because the complexities of modern packaging entail not just a huge waste of resources but also a massive effort for recycling. The modern teabag is a good example. The bag itself and the tea leaves both have to be separated for composting, while the plastic string has to be put in the plastics recycling container and the paper tag at the top into the bin for paper recycling. Some bags even have metal staples that need dealing with. This is a lot of effort for something that did not exist a generation back, when tea drinking was still a very popular habit. Then, tea was sold loose in paper bags and eventually cardboard boxes, and was often being transferred at home into an air-tight tin to keep the tea fresh. The kettle was boiled, the pot warmed and the hot water poured on the loose leaves in the pot. When the tea had been consumed, the pot was taken to the compost heap to be emptied so the tea leaves and any remaining liquid could be disposed of and rot. The paper or cardboard packaging was usually burned on the fire or stove, to complete the disposal of the purchased product.

So, although modern tea bags are sold on the idea of convenience, if the time spent separating the waste for recycling is included in the calculations, tea bags are probably no more convenient than tea leaves. The convenience of the tea bag is based on easier tea preparation and because the user does not have to recycle the resources – the whole tea bag is thrown away after use. And this contributes to the problem of trying to find enough sites for landfill. But of course the tea bag is more profitable than loose tea for the sellers of tea.

The landfill issue raises the question of ways of behaving to avoid gathering packaging that then has to be thrown away.

Some supermarkets, in an effort to reduce packaging waste, have begun to encourage their patrons to bring their own bags, as the plastic supermarket bag is a throw-away item that takes a very long time to bio-degrade. By contrast, about half of all household rubbish will bio-degrade readily in the presence of water and oxygen. Unfortunately, most modern landfills are designed so that virtually nothing bio-degrades, because of the high degree of compaction and the fact the pits are sealed, preventing the entry of air, although there are some landfills where separate sites are created for bio-degradable materials and for things that will not break down. However, this introduces another thing the householder can do and that is to sort rubbish at source because the best place to deal with bio-degradable waste is the home compost heap in the garden or back yard (while providing useful fertiliser for the garden). Other materials, like paper, glass and some plastics, can be recycled and do have a value, but this again depends upon the consumer sorting the rubbish and there being a system for collecting it for recycling.

New Zealanders currently produce 400 kilograms of household waste per person per year.[3] One way local authorities encourage recycling is to provide a limited amount of space, such as a single rubbish bag, for each household each week, and to collect recyclables separately.

Obviously, the best way to deal with the growing waste stream is to reduce consumption and, if you must consume, to refuse excess packaging. The aim, after composting and recycling, should be to have nothing left once all these processes have been undertaken.

Another solution to dealing with packaging is to make things in the home. For previous generations, the home was a place of production, not just a place for consumption. Meals prepared at home and clothes and other items made at home will have less overall environmental impact than their off-the-shelf equivalents. How? Firstly, making clothes at home will mean less transport energy and less packaging will be involved, as there is a substantial difference between moving the basic component parts of cloth and patterns and moving the finished garments that then have to be competitively retailed in air-conditioned shops. Of course, some items (like home projects or toys) use so-called 'waste materials' from around the home and so their production is even less costly to the environment. Secondly, making things also makes clear the effort that goes into the creation of items like clothes, so there may be less temptation to throw them out when fashion moves on. Lastly, making things can be a satisfying hobby that takes up leisure time that might otherwise be spent on that increasingly popular modern pastime of shopping. Making one's hobby consumption will never be good for the environment.

DIGGING FOR VICTORY

In the second world war, the last time that we were exhorted to be resource-efficient, householders were encouraged to grow their own food in 'Dig for Victory' campaigns. Mention has already been made of the productive potential of the space around the house. When statistics were first collected on garden productivity in the 1956 New Zealand Census, it was discovered that although more food was grown in the 'counties' rather than in the 15 designated urban centres, of the households who answered the question (approximately 50 per cent of the population), 40 per cent of households grew at least 25 per cent of their total consumption of vegetables other than potatoes and 31 per cent of households grew at least 25 per cent of their total potato consumption. At that time, the typical New Zealand detached house with its large section was being used in what would now be regarded as a sustainable way.

The problem with the modern food production, retail and packaging industries is that what is being sold as food can really be conceived of as buying oil. Every time we eat, we are all essentially eating oil. Virtually all of the processes in the modern food system are dependent upon this finite resource. One indicator of the non-sustainability of the contemporary food system is the ratio of energy outputs – the energy content of a food product (calories) – to its energy input. The latter is made up of all the energy consumed in producing, processing, packaging and distributing that product. The larger this energy ratio is, the better. The energy ratio (energy output/energy input) in agriculture has decreased from being close to 100 for traditional pre-industrial societies to less than 1 for most of the food products supplied to consumers in industrialised countries, as energy inputs, mainly in the form of fossil fuels, have gradually increased. In modern high-input fruit and vegetable cultivation, the output/ input ratio is between 2 and 0.1 (at an energy ratio of 0.1, one calorie of food energy output requires up to ten calories of energy input). For intensive beef production, the ratio is between 0.1 and 0.03, and the ratio may reach extreme values of 0.002 for winter greenhouse vegetables. All of these ratios refer to the energy consumed to get the food to the farm gate and exclude processing, packaging and distribution, which make the ratios even lower.[4]

Another really important behaviour change is to eat less meat, as vegetable and fruit production usually have higher energy output/input ratios. The figures above suggest a further important behaviour change, and that is – when you do buy food – not to buy food that is out of season. Most recognised cuisines in the world have grown up around the availability of different foods at different times of the year. In the 1940s, New Zealanders had access to two publications, *365 Puddings* and *365 Savoury Suggestions*, both with the subheading 'one for every day of the year', that celebrated the fact that different foods were available at different times of the year. This seasonal variation in the diet has been lost in the seamless availability of produce at the supermarkets, which aim to have the same stock on sale all year, with only the prices reflecting the distance some products have travelled or the additional energy expended in growing them unseasonably. So, even if you cannot grow food, it is possible to be knowledgeable about what foods should be seasonably available, or at the very least be guided by price, and buy accordingly.

'Diet for a small planet'

Diet for a small planet was a recipe book published in 1971 which focused on ways of eating that involved lower consumption of meat and refined foods. The energy inefficiency of our food system can be highlighted by unravelling the supply chains for everyday food products. Researchers at the Swedish Institute for Food and Biotechnology looked at what went into making a bottle of tomato sauce.[4] Processes considered included what went into growing the tomatoes, such as fertilisers and water for irrigation, and those involved in converting the tomatoes to paste, such as fuel used in transportation and energy for cooking. Of more interest were the many different locations where the process happened. The tomatoes were grown and converted to basic paste in Italy, other ingredients were added in Sweden where the product was packaged, the bags for transporting the paste were made in the Netherlands, and the bottles were either made in Sweden or the UK but using materials from Belgium, Denmark, Italy and even the USA and Japan. For example, the screw cap and top were made in Denmark and sent over to Sweden. The study showed more than 52 transport and process stages were involved, and all this for a product that is there only to make something else taste better.[4] This raises the question what is meant by the word 'food', as some foods are essential and some are not.

To highlight the supply-chain problem, it has been suggested in Europe that foods be labelled with their 'food miles' to show what has been involved in their production. Fortunately, the fact that New Zealand still has a substantial agricultural economy means that many of our foods are grown relatively locally. Making a resolution to buy only New Zealand grown food is not just good for the economy, but is also good for the environment.

Choosing food also means choosing a diet. The advantage of growing food at home is that we all tend to like to eat what we grow, and so we will be eating more fruit and vegetables. Not only is this better for our health, it also takes a lot less land and water to gain the necessary calories from fruit and vegetable products, including nuts and beans, than from meat products. Eating less meat, then, is good for the environment. That said the human digestive system is adapted to eat meat, but research shows that we are healthier if we do not eat it every day of the year. The old adage of eating a balanced diet is still the best approach. Meat forms part of most peasant diets, but there is often only a small quantity of meat amongst other ingredients in a meal, and not all meals are meat based. If you are considering reducing your meat intake, now is probably the time to face the fact that our consumption of dairy products is responsible for a good proportion of the meat available – it is derived from the unwanted male offspring of the milking cows.

The productive urban environment
Although not strictly to do with behaving sustainably in the household, the move to urban consolidation ('densification') in New Zealand is relevant to our discussion of growing food at home. Contrary to current trends, the goal needs to be the productive city, not just the compact city as at present. The distinction between the urban and the rural may disappear in a sustainable future, as all environments become places for growing food, amongst other local production. The environment that is built today is going to have to last a long time, so it would be a good idea to future-proof it by ensuring now that there is sufficient space for at least some urban food growing, even if the land is not used for food production until further in the future.

LOCATION, LOCATION, LOCATION: THE IMPACT OF HOUSEHOLD TRAVEL
Real estate agents always say that it is not the house but the location that is important, and this is doubly true for the household trying to behave more sustainably. The further away you live from the places you have to get to, the worse is likely to be your environmental impact, because of the impact of transport.

Household transport accounts for 25 per cent of New Zealand's total CO_2 emissions, and the amount of CO_2 produced by household transport is increasing at a rate far greater than population and GDP growth.'[5] So travel is clearly a problem if we want to be carbon neutral by 2020. Any person walking to work knows the difference in traffic volume between children being at school and children being on holiday. During the holidays, the roads are clear and easy to cross. On school days, they are choked with traffic and, especially around the schools, with cars desperately cruising round trying to find a place to pull in and drop off or pick up children. In the past, neighbourhoods were planned so that all children could walk to school, accompanied when young and on their own as they got older. The walk to school was part of a healthy daily routine and a chance for the children and accompanying adults to talk, perhaps preparing for the day. The walk home was, perhaps, even more important, giving the young a chance to talk through what had happened during the day. For adults, the walk to school and, especially, standing waiting outside the school gates to collect the children was a social occasion. But today, it is hard to foster a sense of community when each family is shut up in its own car. Once again we need to use the walk to school as a bridge between home and the wider community. The walking school bus is a move in the right direction in encouraging a group of children to walk to school under supervision. However, one sensible policy to encourage community and reduce impact from transport would be to go back to the idea of local schools for local people, rather than a 'free for all' approach.

However, how we travel is also important, even if we choose to drive. The following example shows the impact of the type of car you choose to own. Often the car is seen as part of the image you wish to create, but it would be better to make such choices on other grounds, such as the impact on the environment.

The average commuting distance from home to work in Auckland is around 13 kilometres each way. If we assume four weeks' holiday a year, that means the average Aucklander is travelling around 6000 kilometres per year, just to get to work and home again. How much carbon does this travelling produce in a year? John Banks, one-time Mayor of Auckland, used to drive a Bentley. The 6 litre Bentley Continental consumes 26.2 litres of fuel per 100 kilometres on the EU urban drive cycle,[6] so if Mr Banks travelled the average distance in his Bentley, he would create over 3.5 tonnes of CO_2 a year. Every kilometre he drove would produce over 600 grams of CO_2. (Calculation based on 2.31 kilograms of CO_2 per litre of petrol, from National Energy Foundation calculator).[7]

But driving a car does not have to be so damaging. The Toyota Prius hybrid has a fuel consumption of 4.3 litres per 100 kilometres on the European combined cycle[8] and 5.0 litres per 100 kilometres on the urban cycle. A commuter using one of these, or a small diesel car, would emit only around 700 kilograms of CO_2 per year, a modest 116 grams per 100 kilometres. Commuting in a 'typical' car means an annual emission of around 1.25 tonnes, which is more than the weight of the car. Generally, the bigger the engine, the more CO_2 a car emits, as bigger cars use more fuel than smaller ones. And it is important to remember that it is not commuting in a hybrid car that makes the difference – big hybrid SUVs use much more fuel than does an ordinary non-hybrid car. This is another example where 'small is beautiful'.

Of course, commuting between work or school and home is not all the driving that people do. Home-based trips include things like driving to leisure activities and going on holiday in the car. However, rather than driving to the gym, maybe walking the dog or accompanying the children to school are good exercise and have a far lower environmental impact. A car is a good thing to have for those journeys that are too far to walk or for those places that are impossible to get to by public transport, but it should always be the third choice when it comes to deciding how to make the journey from home, not the first choice.

A look at the effect of all the kilometres travelled in the family car gives the following. The AA of New Zealand assumes that an average car travels 14,000 kilometres per year.[9] This amount of driving in a Prius would produce around 1400 kilograms of CO_2, but the same amount of driving in a car having an engine of 2 litres or more would be about 3.5 tonnes. Naturally, choosing not to drive as often would reduce emissions regardless of the type of vehicle.

Ironically, given how today's society prides itself on its progress and technology, modern cars are no more fuel-efficient than the cars of 50 years ago, they just have more cupholders. In the 1961 Mobilgas Economy Run, which was held annually in New Zealand until well into the 1970s and tested production cars on a drive round the country, the winner was a Morris Mini-Minor (the original Mini), which returned a figure of 60.19 miles per gallon.[10] This is 4.69 litres per 100 kilometres. The winner on overall CO_2 emissions in the 2006 EECA Energy Wise Rally was a Smart Fortwo Coupe with a fuel consumption of 4.53 litres per 100 kilometres (62.35 miles per gallon),[11] not much of an improvement on emissions for 45 years of technological progress. So, even though cars have become safer and more reliable, 'progress' in terms of reduced environmental impact has been quite small, and it is even less impressive when you consider that the original Mini was a four-seater, whereas the Smart Coupe has room for only two.

However, to reduce your CO_2 emissions, you could do a lot better than even the 'greenest' car by giving up the car and taking the bus, as a rush-hour bus emits only 20 grams of CO_2 per kilometre (remember that this figure for the Prius was 116 grams per kilometre). So, using the figures for a year's worth of average commuting to and from work, emissions would be only 120 kilograms of CO_2 a year (compared to

700 kilograms for the Prius). Or, you could cycle and produce, at first glance, no emissions. Cycling is a more efficient method of moving people around than walking, and for all the babies out there walking is more efficient than crawling, so give it a go kids! However, the extra food you would need to provide the energy for cycling, compared to travelling by motorised transport, means that the carbon balance for cycling is not zero, because of the energy and emissions associated with producing the food.

When we go on holiday, things become more complicated. Modern aircraft are more fuel efficient than modern cars, in terms of CO_2 per passenger kilometres (one kilometer in a typical petrol car with fuel consumption of 9.7 litres per 100 km emits 0.23 kg of CO_2, but one mile in a short-haul European aircraft emits 0.18 kg of CO_2,[12] but when you fly you tend to go further than when you drive. For example, a round trip from Auckland to London is 36,000 kilometres. Although the airline industry is seeking ways to supply aviation fuels from bio-sources, such as growing and processing algae, at the minute all air travel comes with a heavy environmental tag. Like all things to which we can all too easily become addicted, it should be used with great care.

CONCLUSION

When we look to Europe, we can see that change towards more environmentally friendly behaviours has largely come about because of government policy changes and government incentives. In southern Germany, for instance, many houses have photovoltaic panels and solar heat collectors on their roofs because the government partially subsidises their installation and because national law means that the price paid for the electricity generated and fed back into the grid is several times that paid for electricity generated by conventional means. In the UK, massive policy changes dictating that all new houses be zero-energy users by 2016 are leading to the construction of highly insulated and comfortable houses with very small or no carbon footprints. If New Zealand were ever to have such policies, then the move towards sustainability would happen rapidly. At the moment, change has been left in the hands of us, the ordinary public. What this chapter has tried to show is that there are many things we can do to reduce our carbon footprints that do not incur additional costs, and may even save us money. Reducing the carbon footprint at home is all to do with the life choices that we make. A carbon-neutral future for New Zealand has very much been left up to us.

ENDNOTES
1   http://www.earthsong.org.nz/ (accessed 2 July 2007).
2   The Environmental Choice label, initiated and endorsed by the NZ government, covers paints and other building materials and components   http://www.enviro-choice.org.nz/ (accessed 17 June 2007).
3   http://www.stats.govt.nz/products-and-services/nz-in-the-oecd/environment.htm (accessed 9 July 2007).
4   http://www.resurgence.org/resurgence/issues/jones216.htm (accessed 8 June 2007).
5   http://www.transportco2.org.nz/ (accessed 8 June 2007).
6   http://www.bentleymotors.com/Corporate/display.aspx?infid=233 (accessed 8June 2007).
7   http://www.nef.org.uk/energyadvice/co2calculator.htm (accessed 9 July 2007).
8   http://www.toyota.co.uk/vs2/pdf/PS2_63_spec.pdf (accessed 9 July 2007).
9   http://www.aa.co.nz/motoringimg/Car_running_costs_may per cent2006.pdf (accessed 9 July 2007).
10 Schoenbrunn, R. 2002. *The New Zealand Morris Minor Story*. Wellington: Transpress.
11 Data from results spreadsheet, downloaded from http://www.aa.co.nz/Section?Action=View&Section_id=612 (accessed 8 June 2007).
12 http://www.nef.org.uk/energyadvice/co2calculator.htm (accessed 9 July 2007).

Thinking outside the car:
how we can achieve carbon neutral transport

Julie Anne Genter

Growing up in Los Angeles, Julie Anne Genter was aware at an early age of the disadvantages of automobile dependence. She then had the fortune to go to university in Berkeley, one of the most pedestrian and cycle friendly cities in North America. The contrast was inspiring! Nearly four years living in France further convinced her that cities could be wonderful places if they weren't planned for motor vehicles. The author is currently a University of Auckland Scholar in the Master of Planning Practice programme, and cycles almost every day to get around Auckland.

THE CURRENT SITUATION

'The motor vehicle has been the single greatest factor in forming the pattern of modern urban development. It is probably the greatest mechanical convenience man has yet devised for himself.' (Auckland Regional Planning Authority, 1955[1])

Indeed, motor vehicles were considered such a convenience that we built our economies, communities and lives around them. But where has that got us? Today, we travel farther than ever, sitting in traffic jams at peak hour. Walking or cycling, even in small towns, can be unpleasant and dangerous, owing to noise, exhaust and ugly streetscapes dominated by cars and car parks. The distances between destinations limit the independence of those who cannot drive or do not own a car.

Worst of all, we are still increasing our carbon emissions, at a time when we should be curbing them. Road transport is the fastest growing sector of increasing emissions in New Zealand, and the majority of road transport consists of private vehicles. (See Appendix 1)

THE VISION FOR 2020

By 2020, the transportation system will be vastly different from the car dependence of the late 20th century. You will walk 10 minutes to catch a solar-electric tram to work, stroll by the small local market to do most of your shopping on the way home, and without worry send your children to school on their bikes. Streets and spaces used for car parks and service stations will be productive and recreational spaces for people – not for cumbersome personal motorised transport boxes. Walking, cycling and taking public transport for most of our trips will not only make us fitter, healthier, happier and safer – it will foster thriving local economies and communities.

STRATEGIES TO ACHIEVE THE VISION

To reverse development patterns and allow alternatives to the car to flourish, we need to undo what we have done – in other words, we need to un-subsidise motor vehicle travel. The true cost of motoring – including health, safety, building and maintaining roads, polluting the air and water, and CO_2 emissions – must be internalised. A comprehensive and nuanced policy programme that targets both the demand for motor vehicle travel and the supply of alternatives must be implemented immediately to achieve this vision.

There will be some short-term costs associated with this transformation, but we stand to gain significantly by creating a New Zealand that is not dependent on motor vehicles.

TRANSPORT AND CO_2 EMISSIONS: TRENDS AND CHALLENGES IN NEW ZEALAND

Domestic transport is the largest single sector contributing to CO_2 emissions in New Zealand; it is responsible for over 40 per cent of all CO_2 emissions as of 2006. Only 7 per cent of these are due to domestic air travel, while nearly 90 per cent come from road vehicles. Since 1990, domestic transport has been the fastest growing sector of all greenhouse gas emissions.[2] This is in no small part due to the fact that car ownership has increased, giving New Zealand one of the highest per capita car-ownership rates in the world.[3] Freight transport has also become highly dependent on a large trucking sector. In the 2006 budget, the government announced the biggest investment in road-building we have ever seen.[4] According to a cabinet paper jointly released by the Ministers of Transport and of Climate Change, under current policy and with a business-as-usual scenario, energy use for transport (and thus carbon emissions) will increase by 35 per cent between 2005 and 2030.[5]

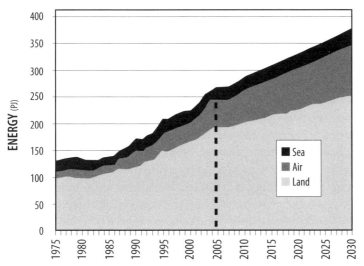

Fig. 1 Projected growth of energy demand under current policy measures from New Zealand's Energy Outlook to 2030.[6]

The high current and projected carbon emissions from road use mean that, in order to achieve the goal of carbon neutrality by 2020, a large part of our effort should be aimed at reducing the dependence on roads in our domestic transport sector. However, there are two closely related assumptions that stand in the way of a comprehensive programme to reduce our transport emissions.

One is the assumption that our current demand for transport, and the system of roads we rely upon, can not change. This view holds that we cannot possibly give up the high level of mobility that we have enjoyed for the last half century, nor the economic development it has facilitated. This assumption has led to much talk of fuel substitution (for example, biofuels or electricity for oil), rather than reduction of the demand for motorised travel. Along these lines, one recent climate change initiative proposed by the government is the uptake of biofuels; a bill has been proposed that mandates by 2012 (the end of the first Kyoto commitment period) that 2.25 per cent of all fuel sold at service stations in New Zealand be in the form of biofuels.[7] The idea seems to be that new technology can entirely replace fossil fuels, and permit us to keep driving everywhere while still achieving carbon neutrality. Of course, a target of less than 3 per cent biofuel use by 2012 is hardly going to be much help in attaining carbon neutrality in New Zealand by 2020. Moreover, there are significant negative environmental impacts that could result from growing enough of a single crop to fill up the fuel tanks of all of our cars and trucks.[8] Even with fuel replacement, there would still be a need for fossil fuels to create the materials that are used in automobile and fuel production in the first place.[9]

Having recognised the difficulties posed by fuel substitution, the government has resigned itself to a second assumption: if the immediate costs of mitigation are greater than the cost of purchasing carbon offsets, or carbon credits, from developing countries in the global carbon market, then it is preferable to purchase offsets.[10] An economic consultant who has close ties to the Ministry for the Environment has publicly expressed the opinion that the costs of significantly reducing the emissions from transport would be too high in the short term, and thus believes that it is more economically advantageous to purchase offsets and try to achieve emission reductions elsewhere.[11]

Both of these assumptions are dangerous. They ignore the fact that our economy and way of life have depended on cheap fossil fuels and an assumed infinite ability of the planet to absorb the pollution created by our use of them. Climate change is just one sign that we have reached the end of that era – our survival and well-being in the future will require a very different approach to transport and trade. Purchasing carbon credits may help invest in better technology somewhere around the globe, but it does nothing to prepare our country for a future without cheap fossil fuels. Neither is it an adequate strategy to become carbon neutral. In fact, if New Zealand becomes a world-leader in achieving carbon neutrality by purchasing carbon offsets, we will find ourselves unable to afford the price of carbon when the rest of the world follows suit. Actual reductions in emissions still must be made somewhere on the planet.

An effective response to the challenge of climate change will, first of all, require the recognition of the hard truth that we need to make big changes. Our economy and way of life will have to adapt to a new environmental reality. The most cost-effective way of responding to this reality is not to purchase off-sets and continue expanding highway capacity, as this does nothing to decouple our economy from carbon and fossil fuels. *The Stern Review of the Economics of Climate Change* has effectively demonstrated that mitigation measures imposed in the short term will cost much less than inaction; 'Strong and early mitigation has a key role to play in limiting the long-run costs of adaptation. Without this, the costs of adaptation will rise dramatically.'[12] Thus, if we expect to benefit from mobility in the future, changing our transport system is not something we can afford to put off.

The reliance on personal motor vehicles as the primary means of transportation (of both people and goods) is known as 'automobile dependence'.[13] The high degree of automobile dependence in New Zealand is directly related to the high carbon emissions that are produced by transport. Moreover, much research has shown that automobile dependence can be extremely costly to society in other ways: poor air and water quality, high costs of roads and parking, high consumer expenditure on transport, obesity and related health problems, motor vehicle accidents, low accessibility for non-drivers and delays due to traffic congestion.[14] By tackling the problem of automobile dependence, we can achieve a significant and long-term reduction in emissions, while also benefiting in other environmental, social and economic ways.

In this chapter, I will present a holistic set of policy measures that we can implement to reduce the carbon emissions that come from domestic land transport.[15] Such measures will also help our society make the transition to a post carbon- and post fossil fuel-dependent economy, and will have significant co-benefits, such as improvement of urban air and water quality, decreased costs from parking and roading, increased economic efficiency, and significant health and community benefits. The approach I will take focuses on an overall gain in transport efficiency – by permanently reducing our automobile dependence. At the end of this chapter, I will also briefly outline immediate changes that people and organisations can make to help reduce their own carbon emissions due to transport.

POLICY MEASURES
Bold action is needed from all levels of government – especially central – as, until there is a comprehensive approach to sustainable travel in place, it will be difficult for many people to lessen their automobile dependence. In fact, I would say that it is unfair for the government to put such a huge responsibility on households and organisations, by asking them to walk, cycle and take public transport, without implementing significant measures to make these modes more practical and economically feasible. It is especially important that legislation and policy measures are applied uniformly, leaving no free-riders or accidental subsidies for behaviour that aggravates the situation (for example, it would be irresponsible to put a tax only on inefficient vehicles that are sold new because it would encourage people to buy older second hand cars that are even more polluting). For legislation and policy changes to be successful, of course, individuals and businesses must support these actions, even if it means paying more for motor vehicle use.

The first step in creating a sustainable transport system in our market-based economy is to internalise the environmental costs created by motor vehicles. Economists describe any costs to society that are not factored into the exchange between the consumer and the supplier as 'externalities'. When a car manufacturer produces an SUV, and I (the consumer) purchase the SUV, the future air pollution and carbon emissions that I will generate while driving the SUV are not factored into the price I pay to purchase the vehicle. Nor are they fully factored into the price I pay to operate the vehicle, even though there is a petrol tax and registration charges associated with operating a vehicle in New Zealand.

Some New Zealanders believe that fuel is already overtaxed (usually because a rise in prices does not fit in their budget). However, despite fuel taxes, we still observe many externalities such as pollution, greenhouse gas emissions and traffic congestion. These are, according to economists, market failures – examples of where the free market does not allocate resources efficiently. We know the cost of fuel and motor vehicles is too low as long as we have the problem of growing carbon emissions, air and water pollution, and traffic congestion. Thus, it is the role of the government to step in and correct these market failures, by employing economic pricing instruments that will internalise these external costs.

Part of the blame for the market failure in New Zealand can be attributed to historic policies, which have effectively subsidised motor transport by publicly funding roads and state highways, removing an import duty on used cars and failing to adopt the carbon tax proposed in 2005.

So how do we internalise the environmental costs of motor vehicles? The short answer is to increase the costs associated with owning and operating motor vehicles and use the money raised to subsidise sustainable forms of transport.[16] However, in order for this to have the best outcome for New Zealand, in terms of continued economic prosperity, quality of life, equity and carbon neutrality, the measures used to internalise the environmental and social costs need to be very nuanced and carefully applied. After all, we have seen that market failures can be exacerbated by a decision to subsidise one form of transport (cars) over all the others.

Carbon-emission fuel tax The first and most obvious measure is a carbon-related fuel tax. Carbon emissions and other air pollutants are related to the amount of fuel a vehicle consumes. Thus, applying a tax to all carbon-emitting fuels that is proportional to the emissions they generate will send a strong signal to the economy. This can have effects like increasing the financial attractiveness of cars that have better fuel economy, discouraging unnecessary trips and encouraging other modes of transport. This measure would affect users in terms of their vehicle usage, as opposed to directly targeting ownership. Thus, those who use a car infrequently and generate fewer carbon emissions pay less. The proceeds from this tax should go primarily to funding sustainable transport alternatives, which I will discuss later in the chapter.

Road pricing and parking levies The second mechanism is also related to vehicle operation. Road pricing and parking levies are measures that put a value on the space required to accommodate motor vehicles during peak hour, both on the roads and in car parks. Thus, it targets directly automobile dependence, not just fossil fuel dependence. These measures are quite efficient as prices can be varied to reflect demand. A road-pricing study was undertaken by the government in 2006 to explore charging in the central area to reduce congestion in Auckland.[17] If road pricing were implemented, driving and parking in the urbanised area of Auckland at peak hours would be very expensive, whereas driving in the middle of the night could be much cheaper. Road pricing could also be used on interurban state highways, and the proceeds used to help fund passenger and freight rail services between cities and towns.

Pay-as-you-drive insurance The third measure is a mandatory pay-as-you-drive insurance scheme.[18] It would increase the cost of operating a vehicle in proportion to how often it is used. This is a much fairer way to pay for insurance, as low users do not subsidise high users (as is the situation now), who are the ones most likely to be involved in accidents and make claims. Theoretical and empirical research has demonstrated that this type of insurance passes on the savings accrued by driving less directly to the consumer.[19]

Fuel efficiency standards Motor vehicles will still be necessary and practical for some trips in the future, so this step addresses the efficiency of the fleet. Regulatory measures that dictate standards for vehicle fuel economy and emissions should be adopted immediately, and should apply to all vehicles. Extra fees that increase as fuel economy decreases (so cars with the lowest fuel efficiency pay the most) can be attached to annual vehicle registration fees, or to Warrant of Fitness applications. Since much of the fleet in New Zealand is old and polluting, many vehicles would be affected. As with road and fuel pricing measures, fees could initially be small and increase dramatically over time. This will create large financial incentives to switch to low emission vehicles where possible, or to give up car ownership altogether.

Car-sharing There are alternatives to car ownership already in many automobile-dependent cities in North America and Australia. Car-share[20] organisations provide members with 24 hour access to vehicles that are parked strategically across urban areas. (Not to be confused with ride-share or carpooling, where several people travel together in a privately owned vehicle.) Members can reserve a car on-line through the

Internet or by phone, and pay for use on an hourly basis. Such schemes permit many urban dwellers who commute by other means, but use a car for short trips on a fairly regular basis, to save money by 'sharing' vehicles among many members. Car-share organisations should benefit from government support, as they would reduce overall vehicle trips and utilise low emissions vehicles. In the long term, they could reduce the need for parking and promote a more compact urban form.

Implementing the policies Most New Zealanders have grown up with cars as the default mode for getting to the shops, work, school and recreation (and even the gym!). So, of course, it can be upsetting when the price of getting to all of these places we have built our lives around suddenly becomes much more expensive. Pricing mechanisms should therefore be introduced gradually, to assist adaptation to new forms of urbanisation and transport. As the transport system has been dependent on private vehicles for many years, it would be unrealistic to think we could make driving prohibitively expensive overnight without causing chaos. However, we can and should act quickly to commence internalising the cost, and increase the various costs on a regular (perhaps monthly or quarterly) basis as alternatives develop. Furthermore, it is crucial that these measures are mandatory and applied uniformly across the country, to ensure that they are effective and that they do not result in market distortions (for example, government needs to avoid motivating people to drive long distances to buy cheaper fuel elsewhere, or to overuse local roads to avoid paying a charge to use the motorway).

Moreover, these measures are not suggested as punishment for automobile dependence! Rather, they are tools that will permit two important routes of change.

1. The true costs of travel will be factored into people's choices.

After an article in *The New Zealand Herald* suggested that the government was considering a 10 cent per litre petrol tax to raise funds for transport projects, including electrification of the suburban rail network, many people wrote angry letters to the paper saying that the proposed tax was not fair, and they could not take public transport because it did not service the areas where they lived and/or worked.[21]

But let us say that you decide to take a job in Manukau City, and you live on the North Shore. You have made those decisions (location of job, home) under a number of constraints. You knew that you would earn a better salary at the job down south, and that the home lifestyle you preferred (and that you could best afford) was on the North Shore. These decisions were made based on assumptions about the cost of travelling between the North Shore and Manukau City. Unfortunately, the cost of driving has not historically reflected its true environmental and social costs, as we have seen. Resisting price measures that would raise this cost to reflect the environmental damage and congestion inflicted by a car journey across town during peak hours, five days a week, amounts to asking the rest of the planet to subsidise your lifestyle choices.

It does not seem very fair to ask this of everyone else, does it? People and businesses have made their decisions about where to locate based around certain assumptions related to the cost of transport – both the monetary cost and travel time. But it is important that the environmental and social costs of travel also be taken into account when choosing where to locate. Obviously, we cannot expect everyone to be able to change his or her jobs or where they live immediately, so I suggest that these pricing measures commence immediately, but at a very low level. They should then be increased steadily to reinforce medium- and long-term decisions to work closer to home, or to live and work closer to public transport. This brings me to the second route of change.

2. The funds raised can greatly increase the coverage and service of public transport, and of rail and shipping for freight delivery.

The funds raised can greatly increase the coverage and service of public transport, and alternatives to freight delivery by trucks, thereby both reducing demand for roads and funding the alternative means of transport at the same time. It is crucial to invest heavily in transport networks that allow people and goods to travel without generating high per-capita carbon emissions. Continued public spending on new roads and increased highway capacity will send out the wrong message; it further subsidises automobile-dependent urban forms that will make reducing our carbon emissions more difficult in the long term. If roads and fuel are appropriately priced, we can curb the demand for new roads, use our existing roading infrastructure more efficiently, and fund the substantial investment in sustainable infrastructure.

Rail is by far the most efficient mode of transporting people and goods in terms of carbon emissions.[22] In particular, high speed electric rail can reduce carbon emissions from transport while promoting a compact, sustainable urban form.[23] These facts have been observed by many government agencies in New Zealand: the Auckland Regional Transport Authority, the Ministry of Energy (now the Ministry of Economic Development) and the Ministry of Transport.[24] However, despite increased spending on rail and passenger transport, over 40 per cent of all the funds allocated through the National Land Transport Account in 2007/2008 will go towards the construction of new highways and local roads. Nearly 40 per cent of the funds will also be spent on maintenance and renewals of existing highways and local roads. Only 12 per cent of all national transport spending is allocated for passenger transport.[25]

Immediate expansion of rail lines, financed in part by the fuel and road pricing mechanisms described above, and in part by the budgeted money already set aside for state highway expansion, will give the country a viable alternative to automobile use. As rail lines are developed and expanded, and the cost of road use increases, more and more people and businesses will adapt and find it more cost-effective to take the train. As more people use them, trains will become more economically viable.

Rail stations could be integrated with bicycle networks, bus routes, and car-share locations to make it easier for people to get to and from the rail lines. In the long term, rail lines have the power to shape urban form, and would thus be accessible to more households and businesses, as they will move to cluster around these lines.

Buses can carry passengers more efficiently than do private cars, but the old diesel buses used around the country may be worse for air quality and the environment than cars. Thus, investment in clean buses that run on alternative fuel is essential. Using smaller mini-buses would be much more environmentally efficient on routes that do not carry many people, and would make the streets safer for cyclists and pedestrians. The government should impose emissions and fuel economy standards, and assist bus transport companies in acquiring replacement vehicles. The cost of replacing the old bus fleet could also be met from the funds that have been set aside for highway expansion.

Investment in alternatives to car-oriented transport systems will help us adapt to climate change, in addition to reducing our emissions. Public transport, especially light and heavy rail, requires much less space than automobiles do. Think of all the space required for cars, including a parking space for each at its final destination. Rail and light rail can carry significantly greater passenger numbers given a narrow amount of space (see Fig. 2). By reducing the need for all that paved over space for cars, we would be able to use the land for other things, like green space or local shops and services, which themselves reduce the need for travel. In addition, decreasing impervious areas like roads and car parks will reduce stormwater run-off. This is important for adapting to climate change, as it has been predicted that rain will be less frequent but much more intense in New Zealand, leading to increases in flooding.[26]

4–5 METRE CORRIDOR WIDTH THROUGH CITY	CAPACITY PER HOUR
An extra lane of motorway	2,400 people
Dedicated busway	12,000 people
Dedicated light rail	15,000–18,000 people
Metro heavy rail (narrow gauge)	20,000–25,000 people

Fig. 2 Corridor comparisons used by the Auckland Regional Transport Authority.[27]

Medium to long term: nearly car-free town centres
It may seem impossible to imagine the cities we have now being transformed into communities where cars are not necessary for most day-to-day trips. However, as soon as we reduce the number of cars in city centres, the more attractive they will become as places to live and work. Motor transport is loud, polluting and dangerous, and makes it less viable to cycle or walk. As fewer vehicles are used in urban areas, cycling and walking will become much more feasible and pleasurable. Once the need for parking and wide roads is diminished, there will be additional space that can be used for dwellings, parks and businesses. Pedestrianising town centres is an excellent way of getting the most economically productive uses from a limited amount of space. In Copenhagen, experience has demonstrated that pedestrianising town centres incrementally brings huge environmental, social and economic benefits to cities.[28] Light hybrid electric vehicles might still be allowed on pedestrian streets for deliveries, moving house and for disabled mobility. The car-free centre of Venice's St Mark's Island demonstrates that freight delivery is entirely possible without motor vehicles.

INDIVIDUAL MEASURES

With the right system of policies in place, it will be much easier and more economically feasible to reduce carbon-generating automobile trips. However, in addition to supporting the strategies outlined above, there are actions that individuals, households and organisations can take immediately, if they have the means.

Reducing your own transport related carbon emissions may be easier said than done, but there are many short- and long-term actions that you can take to reduce your own automobile dependence. The key is to think about the external costs that your household or your organisation incurs by using a car.

- Walk, cycle and take public transport whenever you can. If it costs more to get the bus or the train, remember that you are not paying for the environmental cost of your car trips. If you have the financial means to pay for the bus or the train, and you want to help reduce your carbon footprint, it is worth it to pay more. Even if it is only one day a week that you can give up the car, that is still a 20 per cent improvement on your workday commute. The extra walking and cycling will save time and money that you might otherwise spend at a gym, and the time on a bus or train can give you a chance to read and relax. These alternatives may prove to be far less inconvenient and stressful than sitting in a traffic jam.
- Ride-share (carpool) whenever you can.
- Buy a more efficient vehicle, if you can afford it.
- Get rid of one car, or do not own any at all. Owning a car is expensive, between the costs associated with the Warrant of Fitness, registration, maintenance, parking, insurance and depreciation. If you can use a vehicle much less, it could be worth the money to rent one or take a taxi a few times a month, rather than pay the costs of ownership. There is not yet a car-sharing organisation in New Zealand, but if one is established, be sure to join as a corporate or individual member.
- If you are moving, be sure to consider proximity to public transport, or the ability to walk or cycle for most trips when choosing your location. It may cost more to buy or rent property closer to town centres, but if you include the environmental externalities generated by car trips, it is worth paying higher rent/mortgage to be less dependent on a car. The same is true if you are taking a new job;

if the salary is lower closer to home, it may still be worth it to save the vehicle trips.
- Shop closer to home, even if the prices are higher. It is worth paying a bit more for some items at the local dairy, if it means that you are able to use the car less. Big box retailers are able to sell things more cheaply because transport is under-priced.
- Offset your own emissions with a Kyoto-approved provider. It is not the ideal option, but until you are able to reduce your own automobile dependence, you can invest in better technology somewhere else in the world.

FINAL THOUGHTS

By reducing our automobile dependence, New Zealand can significantly reduce CO_2 emissions. However, bold policy action is required, supported by individuals, households and organisations, to undertake a revolution in the way we conduct our lives. There are many benefits to be gained from such a shift; it has the potential to stimulate our economy, reduce health and accident costs, and improve the quality of our lives. Cars and trucks are practical means of transportation that can get us to remote beaches and transport heavy objects – we will always be able to use them occasionally. But our society and our planet stand to gain everything by making private vehicles the exception to transport, rather than the rule.

ENDNOTES
1 Auckland Regional Planning Authority, 1955 cited in Gunder, M. 2002. Auckland's motorway system: a New Zealand genealogy of imposed automotive progress 1946–66. *Urban Policy and Research 20(2)*: 129–142.
2 Denne, T. 2006. *Achieving emissions reductions at least cost: international lessons for the New Zealand context*. Presentation at Climate Change: The Policy Challenges Symposium, Victoria University Institute of Policy Studies. 6 Oct 2006. Slides available at http://ips.ac.nz/events/completed-activities/Climate%20Change%20Symposium/Proceedings.html. See Figure 1 in Appendix 1.
However, according to the most recent report, transport emissions decreased by 1 per cent in 2005, possibly due to the rise in the price of oil and greater growth in the energy sector in that year. As of 2005, transport emissions are responsible for 18.4 per cent of the total greenhouse gas emissions (42 per cent of CO_2 emissions), of which road transport is responsible for 16.4 per cent (39 per cent of CO_2 emissions). Road transport is the second biggest source of greenhouse gas emissions after methane from ruminant animals. 'NZ greenhouse gases keep rising', *The New Zealand Herald*, 4 May 2007.
3 Offices of the Minister of Transport, Associate Minister of Transport, Minister of Climate Change, Cabinet Business Committee. September 2006. Cabinet Paper. *Climate Change Policy: Overview of Progress Towards Reducing Transport CO_2 Emissions*.

4   Cullen, M. Minister of Finance. 2007. Speech given to New Zealand government Friday, 23 February 2007, 12:58 pm. Speech notes available at http://www.scoop.co.nz/stories/PA0702/S00442.htm (accessed 24 Sept 2007).
5   Offices of the Minister of Transport, Associate Minister of Transport, Minister of Climate Change, Cabinet Business Committee. September 2006. Cabinet Paper. *Climate Change Policy: Overview of Progress Towards Reducing Transport CO_2 Emissions*.
6   The source of Figure 1 is Ministry of Economic Development. 2006. *New Zealand's Energy Outlook to 2030*. Wellington: Ministry of Economic Development. Cited in Offices of the Minister of Transport, Associate Minister of Transport, Minister of Climate Change, Cabinet Business Committee. September 2006. Cabinet Paper. *Climate Change Policy: Overview of Progress Towards Reducing Transport CO_2 Emissions*.
7   Ministry of Transport website: http://www.transport.govt.nz/the-proposed-biofuels-sales-obligation/ (accessed 8 May 2007).
8   Boswell, A. 2007. Biofuels for transport: a dangerous distraction? *Scientists for Global Responsibility Newsletter 33*: 10–13.
9   Litman, T. 2005. Efficient vehicles versus efficient transportation: comparing transportation energy conservation strategies. *Transport Policy 12(2)*: 121–129.
10 Ministry for the Environment. 2007. Climate change solutions: whole of government climate change work programme. In *Purchasing/Obtaining Kyoto Compliant Units* (p. 5). Website: Ministry for the Environment. Available at http://www.mfe.govt.nz/publications/climate/climate-change-solutions-jun06/html/page5.html (accessed 1 May 2007).
11 Denne, T. 2006. 'Achieving emissions reductions at least cost: international lessons for the New Zealand context.' Presentation at *Climate Change: The Policy Challenges Symposium*, Victoria University Institute of Policy Studies. 6 Oct 2006. Slides available at http://ips.ac.nz/events/completed-activities/Climate%20Change%20Symposium/Proceedings.html.See Figure 2 in the Appendix.
12 Stern, N. 2007. *Stern Review of the Economics of Climate Change*. Cambridge: Cambridge University Press. Full Executive Summary, p. xxvii. See http://www.hm-treasury.gov.uk/Independent_Reviews/stern_review_economics_climate_change/sternreview_index.cfm (accessed 26 Sept 2007).
13 Newman, P. and Kenworthy, J. 1999. *Sustainability and Cities: Overcoming Automobile Dependence* Washington, D.C.: Island Press.
14 For a review and comprehensive citation of the relevant research, see the Online Travel Demand Management Encyclopaedia entry on 'Automobile Dependency', available at http://www.vtpi.org/tdm/tdm100.htm. Also see for example Litman, T. & Laube, F. 1998. *Automobile Dependency and Economic Development*, Victoria: Victoria Transportation Policy Institute; Newman, P. & Kenworthy, J. 1999. *Sustainability and cities: overcoming automobile dependence*. Washington, D.C.: Island Press; Litman, T. 2001. *The Costs of Automobile Dependency and the Benefits of Transportation Diversity*, Victoria: Victoria Transport Policy Institute; Litman, T. 2005. Efficient Vehicles Versus Efficient Transportation: Comparing Transportation Energy Conservation Strategies. *Transport Policy*, 12(2): 121-129. Available at www.vtpi.org/cafe.pdf; Litman, T. 2006. Smart Transportation Investments: Re-evaluating the Role of Highway Expansion For Improving Urban Transportation. Victoria: Victoria Transport Policy Institute.
15 All of the measures I present have been thought up and advocated before by many other transport researchers, especially Todd Litman, Peter Newman and Jeffrey Kenworthy.
16 Some may be concerned that using pricing mechanisms will be regressive, and thus greatly disadvantage low-income households. However, there are several means of addressing this issue. Community service card holders and beneficiaries may be charged a lower rate in the short term. In the medium term, a portion of the funds raised through the newly introduced charges could be redirected to assist low-income households to improve their location efficiency, so they do not need to rely on a car.

Low-income households currently drive significantly less than high-income households, so by reducing automobile dependence we will actually increase accessibility for those who already cannot afford to drive.

17 Ministry of Transport. 2006. *Tackling Congestion in Auckland: Auckland Road Pricing Evaluation Study*. Wellington: Ministry of Transport. Executive Summary. The central area scheme and parking levies proposed would raise significant funds after covering the operating costs of the technology used for charging, while reducing congestion. However, as the charges would not apply uniformly across the Auckland Region, the study predicted that there would be an increase in the number of trips outside the area where pricing was to occur. This is why I have argued that any road pricing and parking levies should apply as uniformly as possible across the major regions of the nation.

18 Todd Litman of the Victoria Transport Policy Institute has been a primary researcher who has explored and advocated this measure. See Litman, T. 1997. Distance-Based Vehicle Insurance as a TDM Strategy. *Transportation Quarterly*, Vol. 51, No. 3, Summer 1997, pp. 119-138; updated version at www.vtpi.org/dbvi.pdf.

19 Litman, T. 2007. *Win-Win Emission Reduction Strategies*. Victoria: Victoria Transportation Policy Institute.

20 An excellent summary of the history and evolution of car share organizations and links to further information can be found on the Wikipedia website entry for 'car sharing' at http://en.wikipedia.org/wiki/Carsharing (accessed 03/10/2007).

21 anon. 2007. 'Your views: are you happy with a new regional petrol tax?' *The New Zealand Herald*, 3 May 2007. http://www.nzherald.co.nz/section/466/story.cfm?c_id=466&objectid=10436213 (accessed 24 Sept 2007).

22 'Rail transport is considered the most efficient form of land transport in terms of emissions per unit passenger or freight.' NIWA website. 2007. National Centre for Climate Energy Solutions. Transport Emissions page. available at http://www.niwascience.co.nz/ncces/ghge/transport (accessed 03/10/2007)

23 ARTA (Auckland Regional Transport Authority). 2006. *Rail Development Plan September 2006*. Auckland: ARTA.

24 ARTA *ibid.*; 'Well operated and patronised rail has significantly higher fuel efficiency and lower environmental impacts than road transport. Freight transport by rail or coastal shipping is generally more energy efficient for bulk freight carried over longer distances.' Ministry of Energy. 2004. *Sustainable Energy: Creating a Sustainable Energy System for New Zealand* (discussion paper). Wellington: Ministry of Energy. Available at http://www.med.govt.nz/templates/MultipageDocumentTOC____10124.aspx; 'As the environmental impacts of rail are small compared to those generated by road users, policies aimed at modal shift, including from road to rail, where appropriate, are likely to make the most effective contribution towards achieving environmental sustainability objectives at a regional or national level. Initiatives that encourage freight to transfer from road to rail (where practicable) and encourage greater use of passenger transport, especially urban rail services, will contribute towards achieving a more sustainable transport system.' Ministry of Transport. 2005. *National Rail Strategy to 2015*.

25 Land Transport New Zealand. 2007. *National Land Transport Programme: National Summary*. Available at http://www.ltsa.govt.nz/funding/nltp/2007/national-summary/land-transport-funding.html (accessed 24 Sept 2007).

26 See the Ministry for the Environment *Climate Change Impacts Fact Sheet*, available at http://www.mfe.govt.nz/issues/climate/about/impacts.html (accessed 24 Sept 2007).

27 ARTA (Auckland Regional Transport Authority). 2006. *Rail Development Plan September 2006*. Auckland: ARTA. Available at http://www.arta.co.nz/xxarta/plans-and-policies/ (accessed 24 Sept 2007).

28 Gehl, J. and Gemzøe, L. [trans. Karen Steenhard] 2004. *Public Spaces, Public Life*. 3rd Edition. Copenhagen: Danish Architectural Press & the Royal Danish Academy of Fine Arts, School of Architecture Publishers.

Reducing the carbon burden
from Auckland's transport system

Ning Huang, Penelope Anson and Robert Vale

Ning Huang is a current PhD student at the School of Architecture and Planning, University of Auckland. His research interest is the sustainable development of urban transport. He is presently focusing on analysing the transport issues in the Auckland Region by using the Ecological Footprint method.

Penelope Anson is currently an MA student in Planning Practice at the School of Architecture and Planning, University of Auckland. She has an interest in regional planning and local governance issues.

Robert Vale is a Professorial Research Fellow at Victoria University, Wellington. He has been working on issues related to the sustainability of the built environment for over 30 years. He has received a number of awards for low energy and solar buildings, including the UK's first autonomous house and the first zero-emissions community. He is a member of the United Nations Environment Programme Global 500 Forum for environmental achievement.

THE CURRENT SITUATION

CO_2, primarily from transport and electricity generation, is the quickest growing source of greenhouse emissions in this country. In 2005, the proportion of energy derived CO_2 equivalent emissions that came from transport was 42.4 per cent, the highest proportion in the OECD. These statistics are worrying especially since under the Kyoto Protocol New Zealand is obliged to reduce its emissions of greenhouse gases to 1990 levels over the period 2008–2012. We are a long way from reaching that target; in fact, we are getting further away from it because our emissions from domestic transport have actually been increasing by an average of 3.3 per cent per year since 1990.

The finger can be justifiably pointed at Auckland. Auckland's car ownership is one of the highest in the world. Decades of car-based planning mean that the city lacks the sort of public transport infrastructure that is common in similar-sized cities elsewhere in the world, and this makes its public transport service unattractive, infrequent and underused. The focus on road planning has also made the city unattractive for walking and cycling. Because Auckland is designed only for cars, cars are what Aucklanders want to use.

THE VISION FOR 2020

In 2020, Auckland will have clean efficient public transport powered by renewable energy. Rail will be an important part of this system, linking all modes of transport such as bus and ferry through an integrated timetabling system. People will have a public transport card that will allow them to get on any bus, train, ferry or taxi. Britomart will host electric buses and trains as well as light rail with underground routes around the central city. Auckland will be a great place to walk and cycle, with more people living in and around the central business district and other hubs. People will still use cars for some trips, but cars will be smaller, more efficient and powered largely by biofuels and renewable electricity.

STRATEGIES TO ACHIEVE THE VISION

Transport and land use planning will need to happen together. This means that a shift in Auckland's governance structure is necessary and that its plans, strategies and polices be linked into one regional vision. On the ground, existing walking and cycling infrastructure will need to be upgraded and improvements to the public transport situation made. Renewable energy needs to be applied to all transport modes – solar, wind and hydro -power to generate electricity to power electric trains, buses and cars. Local councils and employers need to encourage and promote carpooling schemes and use travel plans. Developers need to think smarter about how and where they build so that they make it easier to use public transport than at present, and individuals need to support these initiatives by walking, cycling and using public transport whenever possible.

INTRODUCTION

New Zealand has set itself an admirable mandate, to be carbon neutral by 2020. But where does transport fit into this? *New Zealand's Greenhouse Gas Inventory 1990–2004*[1] concludes that CO_2 emissions from the transport sector contributed 18.4 per cent of total greenhouse gas emissions over the 15-year period – transport being the second biggest contributor after methane from ruminant animals. Moreover, CO_2 from transport and electricity generation is the quickest growing source of greenhouse emissions in this country.

In 2005, the proportion of energy-derived CO_2 emissions that came from transport was 42.4 per cent, the highest proportion in the OECD (Organisation for Economic Co-operation and Development). This amounted to 14,005 kilotonnes, 62.2 per cent higher than the 1990 level of 8633 kilotonnes.[2]

As the region with the largest population in New Zealand, Auckland has the most transport travel kilometres and, therefore, the most transport emissions. It is also a city built around the car. In 2004, Auckland had 700 cars per 1000 people and about one-third of car journeys were less than five kilometres.[3] Our high vehicle ownership rate is especially disconcerting when compared to that of other major cities. The Sydney ownership rate is 662 cars per 1000 people; for Canada, it averages at 530; for the USA, at 587[4]; and for Manchester, at 372.[5] The resulting traffic congestion is costing the Auckland economy an estimated NZ$1 billion per year.[6]

What is perhaps more disheartening is that over 50 years ago the opportunity to invest in other forms of transport infrastructure was touted. Auckland nearly had its own extensive urban railway system. The Halcrow Thomas Report in 1950 presented a railway development proposal for the Auckland Metropolitan Area. Again in 1965, the Regional Transit Plan by De Leuw Cather and Company put forward a bus-rail transit system to tackle Auckland's congestion issues.

Sir Dove-Myer Robinson, former Mayor of Auckland, even wrote his own *Robbie's Rapid Railway Plan* in 1969, in which he eagerly supported a rail network.

> Some form of expanded rail services has been discussed, on and off, over the whole of my life, and I have no doubt that a rapid rail transit system will become a reality in Auckland in my time. [7]

Unfortunately, Sir Dove-Meyer's idea has not yet come into action, 15 years since his death.

In this chapter, we suggest ways of bringing Sir Dove-Meyer's idea to fruition. We break down carbon emissions from each mode of transport used in the city, review transport initiatives from comparable cities overseas, examine exciting technologies and set out some concrete ways we can improve Auckland's transport system and get closer to being a carbon-neutral region.

CO_2 EMISSIONS AND THE KYOTO PROTOCOL

Under the Kyoto Protocol, New Zealand is obliged to reduce its emissions of greenhouse gases to 1990 levels over the period 2008–2012. The reality is our CO_2 emissions from domestic transport actually increased over this time – by an average of 3.3 per cent per annum. If this trend continues, by 2012, we will have produced a total of 39,952 kilotonnes of CO_2 emissions in excess of our Kyoto obligations.

Given that 30 per cent of vehicles are based in the Auckland region, meeting the emissions-reduction requirement of the Kyoto Protocol will require a reduction of around 7000 tonnes of CO_2 per day from the 630,000 cars in the region[8], if we assume for the sake of simplicity that the total transport reduction has to come from cars. This amounts to every existing car (assuming a two-litre, petrol engine and 20 per cent occupancy) travelling 48 kilometres less per day (including weekends). This equates to a reduction in travel distance of 17,500 kilometres per year, which is greater than the annual distance driven by the average Kiwi male driver.[9] Remember, this is not achieving carbon neutrality; it is merely getting us back to 1990 levels of carbon emissions.

CO_2 EMISSIONS FOR DIFFERENT MODES OF TRANSPORT

To achieve this sizeable reduction – and more – it is useful to first consider the CO_2 emissions produced from different modes of transport with different occupancy rates. Cars, buses, trains, ferries, motorcycles, bicycles and walking are compared (Table 1). The calculations are based on previous research undertaken by two of the authors.

CO_2 emissions from different transport modes depend on many factors. First, there are technical factors, such as the size of the engine, the type of fuel, the weight of the car, the degree of aerodynamic design. Then, there are social and behavioural factors such as driving style (sensitive versus a heavy-footed one who revs the engine at traffic lights) and the route (urban driving generally uses more fuel than motorway cruising). Third, is occupancy rate. A car carrying five people may be highly efficient in terms of the emissions associated with each person's journey but the same car, all other things being equal, is highly inefficient when there is only one person in it.

MODE	CO_2 EMISSIONS (KG) PER PERSON WITH 100% OCCUPANCY	CO_2 EMISSIONS (KG) PER PERSON WITH AVERAGE OCCUPATION (OCCUPATION RATE IN BRACKET)
Car 1.0-litre petrol car 1.6-litre petrol car 2.0-litre petrol car	0.030 0.040 0.050	0.160 (20%) 0.210 (20%) 0.240 (20%)
Bus 51-passenger city bus	0.016	0.080 (20%)
Train Diesel Inter City 125 Train	0.030	0.150 (20%)
Ferries Quickcat ferry Superflyte ferry Jet Raider ferry	0.042 0.064 0.090	0.210 (20%) 0.320 (20%) 0.450 (20%)
Motorcycle Petrol-motorcycle	0.048	0.068 (70%)
Cycle Pedal cycle	0.017	0.017 (100%)

Table 1 CO_2 emissions from different modes of transport*.

*Details concerning the calculations used to generate Table 1 can be obtained from the authors.

What is clearly evident from the above table is that public transport modes (particularly buses and trains) are generally more CO_2-emission-friendly than private cars, and other statistics reveal that this is true especially during rush hour. Bicycles are also a particularly low emitter, although it is important to account for the extra food that will be consumed by the hungry cyclist! Table 1 also shows that the size of a car's engine matters a lot in determining CO_2 emissions. What are the steps necessary in shifting a city like Auckland away from high-emission options and towards lower ones? First, we'll consider the role that can be played by local government.

LOCAL GOVERNMENT
The greater Auckland region comprises four cities but is governed by three district councils, four city councils and one regional council. Although transport for the region is managed by the Auckland Regional Council, the separate councils do much of their own transportation planning. As everywhere in New Zealand, the local territorial authorities have to plan in accordance with national legislation determined by central government, so to understand the role that local government could play in lowering the carbon burden of Auckland's inefficient transport system, we need to understand the laws, policies and strategies within which the local authorities are operating.

The Resource Management Act 1991 has a number of provisions that direct the activities of local regions. The act and its complementary policies all refer to the need to take a sustainable approach to growth. They advocate the avoidance, mitigation and remediation of any adverse effects that result from development. This development can be directly or indirectly related to transport. Although the laws to enforce carbon-emissions reductions were undermined by the recent Resource Management (Energy and Climate Change) Amendment Act 2004[10], there are still ways in which decision makers at the local government level can ensure new projects encourage public transportation and other low-carbon forms of getting around.

This occurs through regional policy statements, which – in the hierarchy of binding legislation and policy – sit under the Resource Management Act. The *Auckland Regional Policy Statement* for the greater Auckland region contains some concrete carbon-neutral objectives such as the need to create high-density centres and growth corridors to support the public transport system. High-density centres are seen as a means to reduce the emissions of greenhouse gases, curb the loss of rural land and to reduce the dependence on non-renewable energy sources.[11] The policy statement also aims to reduce vehicle emissions by requiring any new traffic-generating activities (construction of shopping malls, hospitals, commercial businesses and residential developments) over a certain size to submit an Integrated Transport Assessment. This assessment must address how the new development will encourage the use of different modes of transport. New developments in the region are to encourage people to walk, cycle and use public transport.

Land-use planning for Auckland is directed by the principles of the Local Government Act 2002. Each city and district council determines how it will plan for future land use and transport but it must consider three strategic planning documents: the *Auckland Regional Growth Strategy (ARGS)*, the *Auckland Regional Land Transport Strategy* and the *Auckland Regional Economic Development Strategy*. These outline the region's development for the next 30 to 40 years. Like the Regional Policy Statement, these strategies have objectives promoting carbon neutrality and are trying to get land-use and transport planning to coincide.

The ARGS plans to provide for more dense development and mixed-use communities. 'Mixed use' means combining different types of land uses together – employment, education, shopping and recreation can then all be accessed locally and easily. The ARGS emphasises that, in the next 50 years, most future growth (70 per cent of new houses) will take place within the existing metropolitan area and that this growth should be concentrated around current town centres and major public transport routes, in line with the overall aim of increasing density within the city. The ARGS also envisages that about 30 per cent of future growth would need to be accommodated in some new 'greenfield' (land that has not previously been built on) developments on the outskirts of the region. This would prompt the need for new centres and provide opportunities to develop mixed-use communities.

As if there were not enough plans and strategies already, each city and district council must produce a Long Term Council Community Plan (LTCCP). These are based on a ten-year planning horizon. They focus on transport planning at the smaller scale but can have a fundamental impact on carbon emissions through the policies therein on parking services and enforcement, cycle ways, traffic control mechanisms, travel demand management plans, pedestrian services (footpaths) and rapid transport services.

Central government is well aware of the difficulties facing Auckland in establishing an effective transport system. In 2004, it introduced an act specifically for the region (the Local Government (Auckland) Amendment Act), which aims to tie these strategies, the policy statement, the district plans and the LTCCPs together. Whether this new act will achieve this end, and facilitate transport planning, is yet to be seen. A better approach to improving Auckland's transport system could be through a more strategic approach – a single strategic planning framework for the metropolitan region. Adopting this approach would mean integrating all planning documents around a coherent single vision and one time frame. If we are to reduce our dependence on private transport and get more people using public transport, focused leadership and coherent, simplified legislation is needed.

The councils' role in new developments
In the meantime, the Integrated Transport Assessments required for new developments are an important way the region's local councils can make a difference to CO_2 emissions produced from transport across Auckland. Property developers for any new residential and commercial projects beyond a certain size have to submit a draft travel proposal in their resource consent application. The proposal must outline how the project intends to encourage the uptake of sustainable transport options. There is plenty of scope for developers to be creative in the ways they will encourage public transport use. Cycle facilities, such as secure bike parking, could be designed into any new development. The main access to buildings could be positioned close to public bus stops and taxi ranks. Separating car-parking facilities from buildings is another simple yet effective planning provision. This form of 'structured inconvenience' means that car drivers are likely to think twice before deciding how to get to their local shopping centre or work place.[12]

There are other ways councils can excite developers and the public about better transport options. Builders and developers could be offered bonuses for any new development built within a radius of 500 metres to one kilometre of a major public transport node. Incentives could be given to convert land currently utilised for car parks to a different type of land use. Conversion of parking spaces into permeable surfaces (for example, involving landscaping) could be made a viable option for commercial and residential property owners. Tradeoffs with some other development controls, like maximum floor area ratios or reducing the amount of development contribution payable, could be used as motivators for converting car parking areas, as could a rebate scheme based on the size of land converted and its new use.

Auckland City Council requires entertainment facilities (cinemas, clubs and gyms) to provide one car parking space for every three people while offices have to provide one space for every 40 square metres of gross floor area. These regulations do not take into consideration the location of the activity relative to public transport. Revising them with that in mind (for example, providing fewer car parking spaces if close to public transport) would be an effective way to reduce the number of cars on the road and encourage uptake of public transport options. And, while only a remedial strategy, reducing the required 4.8-metre width of garages could prompt a rethink in car-purchasing decisions. (The required width has increased over time to allow space for larger (less fuel-efficient) vehicles such as four-wheel-drive vehicles.) Similarly, reducing the number of car parking spaces required for each residential unit from two to one could also decrease our currently high use of private transport.

Local councils can also take a 'soft' approach in promoting carbon-friendly transport modes – this means providing support and education rather than changing or introducing a new regulation. Travel plans are a good example of something that councils can promote. A travel plan is a package of measures that widens the range of sustainable transport choices available to an organisation. The organisation can be a school, university, club or employer. The plans can include: maps of the local area to help walkers and cyclists; better information about public transport; a carpooling scheme; efficient car park management; an intranet travel site for ride-sharing; personalised journey planning services and cycle facilities (secure bike parks and showers for those commuting by bike). Employers could go one step further and consider offering subsidised discounts for workers using public transport or give priority allocation of staff car parks to those who carpool. The council in Darlington, UK, for example, is promoting 'The Local Motion'. The programme incorporates a website, distributes information in local schools, retailers and community centres, and generally offers support and ideas for getting residents to use less energy-dependent means of transport. It coordinates carpooling, provides accurate public transport timetabling information and educates members of the public on the benefits of leaving their cars at home. Tag-lines such as 'Did you know that car passengers in slow-moving traffic face pollution levels in the car three times higher than those experienced by pedestrians or cyclists?' illustrate how the co-benefits of using your feet to get around (see next section) are being used to promote sustainable transport options.[13]

WALKING AND CYCLING

How else can we improve Auckland's transport system and make a dent in our carbon emissions rate? Auckland is typical of most New Zealand cities in that it has not been designed to encourage walking and cycling. For example, traffic lights allow little time for pedestrians to cross and crossings are not provided between bus stops on opposite sides of a busy main road. Often pedestrians are forced to wait for long periods in the rain at automatic crossings, while people in dry and heated cars just cruise on by. Cycle lanes are few and far between, stop abruptly and are shared with buses.

Cities overseas have made impressive efforts to improve the environment for pedestrians and cyclists. Their authorities recognise that the walking and cycling experience must be better than the driving one if people are going to switch to lower carbon-emitting modes of transport. The planners of Aarhus, in Denmark, instigated considerable changes between 1993 and 1995 with the aim of creating a better environment for pedestrians. A major measure was to reconstruct some city centre areas to exclude all motorised traffic. This improved safety and access for pedestrians and resulted in almost a 60 per cent decrease in traffic in the central areas. In Enschede, in the Netherlands, the existing cycle network has continually been extended and improved since 1978. In 1995, 35.6 per cent of all trips were made by bike there. In the same year, the Dutch average was 27.8 per cent.[14] In contrast, even as recently as 2001, the New Zealand figure was a measly 2 per cent.[15]

Cycle lanes play an important transport role in some developing countries. In China, cycle lanes are everywhere and most people – no matter what their jobs are – go to work by bike. Interestingly, this is one reason why the transport sector in these countries does not contribute such a high proportion to their CO_2 emissions.

Fortunately, some promising steps are being taken in Auckland. A NZ$30 million project for upgrading Auckland's Queen Street commenced in 2006. The project includes widening the footpath, improving the pedestrian crossings (new signals show how long it will before the light changes back to red), adding more seats and trees and installing a cycle and bus lane in what was once a car lane. This type of central city street upgrade has been occurring in China and across Europe for many years.

Auckland could take the Queen Street upgrade one step further. More Aucklanders could be encouraged to use their feet by converting strategic streets and important tourist and archaeological features into car-free zones. With careful traffic management of the surrounding road networks, streets like Karangahape Road, Queen Street, Grafton Bridge and the northern end of Ponsonby Road could easily be pedestrian-only zones. City beaches and waterfronts (Mission Bay, St Heliers, Devonport and Browns Bay) could be 'carbon-free' zones on weekends and important volcanic cones could have restrictions that ban cars from their summit roads.

PUBLIC TRANSPORT

In 1955, 58 per cent of all journeys made by Aucklanders were taken on public transport. By 1963, this had reduced to 22 per cent, by 1991 to 7.6 per cent and by 2005 to 7 per cent. This means that today Auckland's public transport patronage is among the lowest in the world – lower than any big city in Australia or Canada, and even lower than almost all cities of the USA.[16]

To be fair, we should note that while public transport is sometimes a feasible option for the region's residents, it is not always a good experience. Buses and trains can be infrequent, particularly outside the rush hour. Many bus stops do not have shelters, so the negative effect of a long wait can be compounded if you get soaked by one of Auckland's unpredictable rain showers. Bus stops and railway station upgrades have sometimes been done with materials inappropriate for maintenance – graffiti-proof or easy to maintain materials should be the rule for all public transport shelters. No one likes to wait in a dirty, vandalised shelter.

However, after so many years of the focus being entirely to support car travel, there are now plans for an upgraded and efficient passenger transport system and for the growth of public transport around main transport nodes and along bus or rail corridors. The *Auckland Regional Growth Strategy* explicitly outlines the nodes along the western, eastern and southern railway lines such as Henderson, New Lynn, Otahuhu and Papakura. Bus corridors include Great South Road, Great North Road and Dominion Road. But in spite of these promising policies, other changes are urgently needed to supplement these efforts. A conclusion from an on-line debate on *The New Zealand Herald*[17] site shows clearly that people will not choose public transport unless the quality is improved. The survey relating to the on-line debate entitled 'Your Views: Will you abandon your car?' reported that although the majority of respondents said they support public transport, they will not use public transport modes until they become safer, more reliable, more cost-effective, more comfortable, more accessible and cover more areas.

The survey and debate point to vital drawbacks in the public transport infrastructure of Auckland. For instance, the bus service network in the region is basically radial, so that all the buses head to the city centre and then out again. Bus routes that link different suburban areas are rare. Also, unlike its peer cities with a similar population such as Perth in Australia and Leipzig in Germany, railways in Auckland have been neglected. Even Wellington, which is much smaller than Auckland, has a functional railway network! The Auckland City Council[18] once stated that light rail was a panacea for Auckland's transport problems – and it may well still have a vital role to play.

Renewable energy and public transport

One way to reduce transport emissions in the Auckland region and throughout the country is to use renewable forms of energy to fuel motorised vehicles. New Zealand already has one of the highest rates of renewable energy supply among developed countries (29 per cent of consumer energy, compared with 6 per cent for Australia and the USA and 25 per cent for Sweden).[19] To apply renewable energy to transport, one strategy would be to make more use of electricity (generated from renewable resources) for powering some transport modes.

When Al Gore's climate change film *An Inconvenient Truth*[20] was released, a large billboard at Wellington airport showed a picture of a trolley bus with the slogan 'A convenient truth – Wellington's electric buses'. Electric buses and trains would be an excellent part of the emission-reducing solution for Auckland as well. If the electricity powering them comes from renewable sources, they are 'carbon-zero'.

Integration of all these transport initiatives is sorely needed. Britomart Transport Centre, in central Auckland, could be retrofitted to host electric buses, trains, even light rail or an underground. Public transport should be frequent and fast. A multiple-mode public transport card would allow people to use any public transport vehicle – whether bus, train, ferry or light rail – making journeys easier for all users. Timetables between the different modes should also be coordinated to ensure minimal waiting times at transfer points. This is where a regional focus for transport planning (rather than the current separation between cities and districts) becomes even more important.

IS THERE ANY FUTURE FOR CARS?

Improving the greater Auckland region's walkability, and encouraging cycling and public transport use are absolutely critical to a regional approach to transport. However, cars can still play a role in a carbon-neutral future.

Carpooling

Carpooling has an important role in a less carbon-intensive transport system in the city. The main disadvantage facing carpooling is the lack of public acceptability because of its restrictions on the drivers' and passengers' freedom. So, other benefits need to be provided. The presence of HOV (High Occupancy Vehicle) priority lanes on major roads gives car drivers an incentive to share their trips, a method which has been applied in practice for some years in North America.

The Chief Executive of the Ministry of Transport, Dr Robin Dunlop, has commented on Auckland HOV lanes:

> The use of carpooling can be very effective, but if the right incentives are not in place most car drivers will not consider this option. By way of example, it is quite common for young children to wait near the start of a HOV lane for drivers to pick them up, for example, in Indonesia. For a small charge, they ride in the vehicle and jump out at the end of the HOV lane … By creating bus-only lanes and/or bus priority at traffic lights, combined HOV and bus lanes or just HOV lanes, congestion will be reduced. [21]

As well as the fuel savings and emissions reductions resulting from greater car sharing, Auckland Regional Council also says that carpooling combined with a series of parking prohibitions, turning restrictions, access control and bus/HOV lanes similar to Red Routes in London[22] could improve travel times by 5-10 per cent.[23]

A study in Rome suggests that media campaigns highlighting the environmental benefits of carpooling, together with an organisation that matches up potential poolers according to their professional, social and even musical interests (relevant because carpoolers also share a car's audio system), can encourage people to participate in such schemes.[24]

Car size and energy source

Driving cars that are smaller (in terms of engine size) can directly impact CO_2 emissions. As shown in Table 1, a 1.0 litre petrol car has three-fifths the fuel consumption of a 2.0 litre car and so it releases three-fifths the CO_2 emissions. A number of central government agencies are already leading the way by reducing the size of the cars in their fleets. For example, the Accident Compensation Corporation, the Inland Revenue Department and the Ministry of Social Development have all invested in vehicles with greater fuel efficiency. More than 93 per cent of their combined fleet is cars with engines less than two litres.[25] Small cars have always been preferred in some countries (like Japan), and for good reason. Small cars are cheaper to buy and cheaper to run, and are easier to park than big cars.

Using electricity to power cars is beneficial not just in the ways outlined earlier (that is, less harmful to the environment). The following enthusiastic description lists some of the advantages.

> ... now it is possible for a vehicle to complete 100 miles [160 kilometres] or more ... upon a single battery charge ... The advantages of the electric vehicle are so pronounced as to impress everyone. It offers smoother running; vibration is absent; control is easier ... it is cheaper to maintain ... it is cleaner to handle, and when standing still, it does not eat up fuel, the current being switched off during such periods of inactivity. It possesses all of the mobility of the petrol-driven car with none of its shortcomings ... [27]

What is shocking is that this was written 90 years ago, and that even then you could get an electric car that could travel 160 kilometres before requiring recharging – more than enough for most daily travel.

There are a few electric cars in New Zealand. Neville Watkin, the founder of Watkin Technology in Wellington, has studied the electric car for more than 25 years and even built some of his own.[28] An electric commuter car, using sophisticated direct current motors and lead acid batteries was built in prototype form by Heron Motors in Rotorua.[29] But more support from the government and from drivers is needed if the adoption of the electric car is to become widespread – though the cost of it is still somewhat high (estimated at NZ$30,000, which includes the car and enough solar panels to charge it). Inspiration can be found in an electric vehicle named the 'Reva', which was launched in India in 2001. It is now sold in many cities in India and exported to other countries. On a single charge, the Reva can take two persons a full 80 kilometres.[30] Some of the world's largest automobile companies have explored electric vehicles for several years – for example, General Motors' EV1 and Ford's Think. Importing more electric cars instead of second-hand cars would make this type of vehicle at least an option for Auckland drivers and would certainly be helpful in creating a carbon-neutral New Zealand.

Besides the electrically-charged battery car, other technologies such as that for hybrid electric cars and fuel cell-driven vehicles have been developed by many motor companies. However, these new technologies are not the dream solutions they appear to be. An efficient hybrid car can achieve fuel consumption of less than 4 litres per 100 kilometres, but a hybrid SUV will always use more fuel than a conventional small-engined hatchback. In the same way, fuel cells could be another way towards carbon neutrality, but only if their hydrogen fuel is not made from oil!

Biofuel has some potential. The government has stated it will require an increasing proportion of cleaner-burning biofuels to be sold in order to cut greenhouse gas emissions.[31] This sales obligation starts in April 2008 and would require 3.4 per cent of the total fuel sold by oil companies to be biofuel by 2012. Biofuels currently sold in New Zealand are bio-diesel and bio-ethanol. The former is made from vegetable oils and tallow (animal fats), while the latter is produced from fermented sugars and starches, such as maize, dairy whey and – in the future– wood waste. Care would need to be taken to ensure that these fuels were made from crops grown only in New Zealand (to avoid increasing our carbon emissions from transporting them here) and that their energy consumption and carbon emissions were balanced positively. If the goal is to reduce global CO_2 emissions, it is no use making biofuels from palm oil grown in plantations that were once tropical rainforests.[32]

SUMMARY

Huge improvements could be made to Auckland's transport structure. We suggest:

- Upgrade the existing the walking and cycling infrastructure (especially in the central business district) to encourage more walking and cycling. This type of infrastructure should be mandatory in any new mixed-use communities.
- Focus on transport planning at the regional level by integrating all the existing polices, strategies and plans. Consideration must be given to land-use planning at the same time.
- Improve the poor public transport situation (facilities, vehicles and maintenance), and electrify the current diesel trains. Make buses and railways safer, more reliable, more comfortable, accessible and cost-effective and make sure their networks cover more parts of the region.
- Apply renewable energy to transport modes, to utilise solar power, wind power and hydro-power to generate electricity to power electric trains, buses and cars.
- Promote and sell more biofuels produced internally within New Zealand.
- Encourage and promote carpooling schemes and public transport use through travel plans.
- Encourage increased sales of small engine vehicles by some preferential financial policies such as reduced licensing fees.

At the same time, individuals can do the following:

- Think whether you really need to make that trip at all.
- Walk or cycle whenever possible; it will make you fitter, reduce emissions and save you money.
- Take the bus or the train to work, school and university.
- If you want to buy a car, choose the smallest one that meets your needs. It will cost you less to buy and a lot less to run.

The current situation, one where domestic transport emissions are increasing by an average of 3.3 per cent per annum, can be turned around. Integrating the region's policies and strategies into one vision that deals with land and transport planning at the same time is an important strategic goal. At an operational level, we have suggested key ways to reduce CO_2 emissions. Walking and cycling, as low carbon emitting modes of transport, can be encouraged through infrastructure upgrades. Public transport investment and better car purchasing decisions will also curb our emissions rate. We have looked at cities overseas that have implemented these suggestions and the suggestions are working. These cities' carbon-emission rates from domestic transport are relatively lower than those of Auckland. Auckland needs to get inspired and take action. Addressing the region's transport system is crucial to becoming carbon neutral by 2020.

ENDNOTES
1. Ministry for the Environment. 2005. *New Zealand's Greenhouse Gas Inventory 1990–2004*. Wellington: Ministry for the Environment. Available on http://www.mfe.govt.nz/publications/climate/nir-apr06/html/index.html (accessed 24 Sept 2007).
2. Ministry of Economic Development. 2006. *Revised New Zealand Energy Greenhouse Gas Emissions 1990–2005*. Wellington: Ministry of Economic Development.
3. O'Fallon, C. (2004). *Pinnacle Research & Policy: Short trips and chaining*. Available on www.pinnacleresearch.co.nz (accessed 27 Aug 2007).
4. Ministry of Transport. 2003. *Submission to Ministerial Inquiry into Public Passenger Transport: Sustainable Transport Meeting Community Needs*. Available on http://www.transport.nsw.gov.au/inquiries/parry-MoT-submission.pdf (accessed 27 Aug 2007).
5. Commission for Integrated Transport. (2001). *Study of European best practice in the delivery of integrated transport: report on stage 2 case studies*. Available on http://www.cfit.gov.uk/docs/2001/ebp/ebp/stage2/03.htm (accessed 27 Aug. 2007).
6. New Zealand Business Council for Sustainable Development. (n.d.). *Beating traffic congestion: new approaches to road pricing*. Available on http://www.nzbcsd.org.nz/economicincentives/content.asp?id=343 (accessed 27 Aug. 07).
7. Robinson, D.M. 1969. *Passenger transport in Auckland : a report on overseas investigations of rapid-rail transport*. Auckland: Auckland Regional Authority.
8. Auckland Regional Council (ARC). 2004. *The Auckland Region: Facts and Figures*. Auckland: Auckland Regional Council.
9. Ministry of Transport. 2007. *Household Travel Survey*. Wellington: Ministry of Transport.
10. Ministry for the Environment. 2004. *Resource Management (Energy and Climate Change) Amendment Act 2004*.
11. Auckland Regional Council (ARC). 2005. *Auckland Regional Policy Statement Proposed Plan Change 6: Giving Effect to the Regional Growth Concept and Intergrading Landuse and Transport*. Auckland: Auckland Regional Council.
12. Organisation for Economic Co-operation and Development (OCED). 2002. *Policy Instruments for Achieving Environmentally Sustainable Transport*. Paris: OECD Environment Directorate.
13. 'Do the local motion' website: http://www.dothelocalmotion.co.uk (accessed 22 Aug 2007).
14. DANTE Consortium. 1997. *City strategies and measurement of their impact on avoiding the*

need to travel (Appendix 12). DANTE project Deliverable 2A. Report prepared for EU DG VII. PLS Consult, Aarhus, Denmark.
15 Statistics New Zealand. 2001. Census 2001: *How do people get to work activity*. available on http://www.stats.govt.nz/NR/rdonlyres/537B3A30-2EC3-41C5-A3C9-D42B49486F27/0/Travellingtowork.xls (accessed 2 Oct 2007)
16 Kenworthy, J. et al. 1999. *An International Sourcebook of Automobile Dependence in Cities 1960–1990*. Boulder: University Press of Colorado.
17 *The New Zealand Herald*, Tuesday, April 24 2007. Survey results available on http://www.nzherald.co.nz/feature/story.cfm?c_id=1501154&objectid=10435680&pnum=0 (accessed 2 Oct 2007).
18 Auckland City Council (ACC). 1998. *Central Area Transport Plan: Draft for Consultation*. Auckland: Transportation Planning Division.
19 Department of Prime Minister and Cabinet. 2003. *Sustainable Development for New Zealand. Programme for Action*. available on http://www.beehive.govt.nz/hobbs/30199-med-susined-developm.pdf (accessed 22 May 2007).
20 *An Inconvenient Truth* (motion picture). 2006. Paramount Classics and Participant Productions.
21 Dunlop, R. 1998. *Highway Traffic Growth and Passenger Transport*. Presentation at the 1998 annual conference of the Institution of Professional Engineers of New Zealand. Auckland.
22 The Red Route is a network of 580km of London's roads, which carry 35 per cent of the city's traffic. Free-flowing traffic on these roads is essential to keep London moving - parking on the red route causes jams and clogs up the system. TfL is responsible for traffic enforcement along the route, using a combination of traffic wardens, community support officers and an extensive network of CCTV cameras. More information is available at http://www.tfl.gov.uk/roadusers/finesandregulations/949.aspx
23 Auckland Regional Council (ARC). 2003. *Auckland Transport Strategy and Funding Project*. Auckland: Auckland Regional Council
24 Banister, D. and Marshall, S. 2000. *Encouraging Transport Alternatives: Good Practice in Reducing Travel*. London: The Stationery Office.
25 Parker, D. 2007. *Public Service takes carbon neutral lead*. See the official website of the New Zealand government http://www.beehive.govt.nz/ViewDocument.aspx?DocumentID=28361 (accessed 24 May 2007).
26 Produced by Robert Vale from data provided in AGO. 2002. 2001–2002 Fuel Consumption Guide. Canberra: Australian Greenhouse Office.
27 Talbot, F. 1916. *All About Inventions and Discoveries: the Romance of Modern Scientific and Mechanical Achievements*. London: Cassell and Co.
28 *Watkin Technology specialises in solar power and sustainable transport*. See http://www.watkintechnology.com/ (accessed 22 May 2007).
29 Vale, B. and Vale, R. 1998. *Sustainable Transport in the Twenty-first Century*. Lecture to Engineers for Social Responsibility at University of Auckland on 26 November 1998.
30 For more information about the Reva, see http://www.revaindia.com/ (accessed 22 May 2007).
31 New Zealand Labour. 2007. *Government requires Biofuels sales*. Available on http://www.labour.org.nz/news/hot_topics/HT-070213/ht4328907/index.html (accessed 2 Oct. 2007).
32 See, for example, http://www.newscientist.com/blog/environment/2007/07/freds-footprint-perils-of-palm-oil.html (accessed 29 July 2007).

How 'Hobson Mall'
became climate and people friendly

Maggie Lawton and Robert Vale

Maggie Lawton is a biochemist and chemist by training. Until recently, Maggie was a member of the Executive Management Team of Landcare Research where her research focus included rural and urban development and climate change. She oversaw the development and construction of the Manaaki Whenua Landcare Research building on the Tamaki Campus, University of Auckland, which won national acclaim for its environmental features. Her environmental consulting company has recently been researching and implementing water management approaches in the urban environment. She is also working with businesses and communities on sustainable development.

Robert Vale is a Professorial Research Fellow at Victoria University, Wellington and a Senior Researcher at Manaaki Whenua Landcare Research. He specialises in the design of zero-energy and autonomous buildings.

THE CURRENT SITUATION
'Shop until we drop' has become a reality in New Zealand, as have its consequent heavy debt levels. Shopping has become a recreational activity, especially at weekends when malls are swarming with people purchasing goods that they did not know they needed. The number and size of malls is growing steadily, with huge parking areas showing they are designed with car transport in mind. Goods are cheap because they have been produced by cheap labour in Asian countries and because of the ready availability of oil-based products that form the basis of much of what we buy.
In terms of environmental impact, it is not just a mall's buildings and concrete surrounds that cause the concern, it is the volume and type of products malls promote, the distances those products travel to market and the waste produced in their production, processing, packaging and transport. Each aspect of the life cycle of these products contributes to the increasing level of greenhouse gases and related climate change consequences.

THE VISION FOR 2020
All malls will be either constructed or retrofitted using low-impact urban design principles. They will generate their own energy, catch and treat water on-site and will be situated by public transport hubs, thereby negating the necessity for large car parks. The latter will have, in the main, been turned into children's play areas, community buildings and gardens. Malls will be carbon neutral in their operations and will provide a dramatically different mix of produce and services compared to the malls of 2007. Their suppliers will include a high proportion of local food producers and manufacturers; they will be providing a central focus for community-based activities.

STRATEGIES TO ACHIEVE THE VISION
For malls to survive the impact of rising oil prices, and consequent decrease in the availability of cheap products, their owners and operators will have to adapt to a lower cost structure with better environmental and social outcomes. Incentives to encourage owner/operators to support local producers and service providers are needed, as are motivators to reduce the waste produced by malls.

INTRODUCTION

Despite New Zealand's relatively low population base of four million, malls are continuing to be built. Shopping has become a national pastime, which raises the question: What did people do on Sundays before malls existed? The focus of this chapter, the fictional Hobson Mall, built in early 2007, is typical of the many new malls that have sprouted up in our towns and cities in the last few years. We use Hobson Mall here to show both how malls were designed in 2007, and how they might change in the future in order to become more sustainable.

Although malls are only one of many contributors to climate change, in many ways they epitomise the throw-away society we have become. They are strongly car centred and seldom linked to good public transport systems. Their operations are energy intensive and stormwater pours off their impervious concrete and bitumen surfaces, contributing to pollution and flooding problems already exacerbated by our ageing urban infrastructure. Their stores sell a high percentage of oil-based products that cause the emission of carbon dioxide in their manufacture, not to mention emissions produced in transporting these goods from Asia and elsewhere. They also produce waste, contributing to landfills and methane production. And malls in New Zealand are getting bigger and bigger, some of them imitating mini-towns, swallowing the customers of the nearest city's retail heart. The social changes and costs that the malls produce are clear, the individual retailer in the central business district does not have the same pulling power, and family retailers are now relegated to the corner dairy and not much else.

Because New Zealand does not have a strong manufacturing base, it needs to import a wide range of products, which artificially reduces its carbon footprint and ensures the pollution generated by the manufacturing process stays elsewhere. Despite our manufactured goods being produced off-shore, our carbon footprint is still relatively high, we have one of the lowest rates of saving, mostly living beyond our means, and we are individually and nationally in debt.[1]

How sustainable are the malls dominating our urban centres in 2007? Will their success be relatively short-lived? Many of the pillars on which they are built are looking less secure in the future. Pressures will include reduced access to cheap products, increased transport costs and the increasing concerns about climate change that will eventually impact on customer behaviour.

So what can mall owners and operators do? Should they aim to make big profits now and then go swiftly to financial ruin or could they be transformed into socially and environmentally aware centres of community services, which could be viable in 2020? It is unlikely that the dire but vague warnings about climate change or the tentative attempts by government to curb our carbon emissions will have much impact on the behaviour of mall owners and operators; that approach has resulted in little, if any, change to date.

The crunch is likely to come after 2010, the date favoured by many analysts for when the world's use of oil will outstrip its rate of production – the time of peak oil. While there continues to be debate on the actual year and the speed with which peak oil will impact us, it is widely accepted that oil is a finite resource and that there are basic laws that describe the depletion of any finite resource: production starts at zero; production then rises to a peak, which can never be surpassed; then production declines until the resource is depleted.

These simple rules were first described in the 1950s by Dr M. King Hubbert, and apply to any non-renewable resource, including the depletion of the world's petroleum resources.[2] Increased demand and shrinking supplies mean that oil will become scarcer and prices will rise, transforming the current marketplace. While many of us know that the time of peak oil is coming, now – in 2007 – its consequences still seem unreal. The idea is so foreign – that life as we know it could be so changed, so constrained, as a resource we have come to depend on dwindles to a highly rationed and limited supply.

THE WORLD IN TRANSITION

So what will the impacts of oil shortages and the consequences of climate change mean for the future of our infrastructure such as the fictional Hobson Mall? Malls are huge components of a country's infrastructure, with millions of dollars in capital investment. Suddenly their days of prosperity will be over; like any organism they will have to adapt or die. In the rest of this chapter, we explore a possible scenario for the decade from 2010.

Cheap imports into New Zealand from the manufacturing economies of Asia will dwindle as these economies begin to falter. China may continue to spew coal emissions into the atmosphere for some time and use all the fossil fuels they can convert to keep their transport and factories going, but eventually they will have to recognise that the days of exporting large amounts of cheap goods are numbered. Air pollution is even now killing people in industrial centres and making life unpleasant for much of the population. People in China are already turning their ingenuity towards solving China's internal environmental problems and are beginning to make a serious attempt to develop sustainable ways of living within their own resource availability and ecological footprint. The reduction in production of cheap goods from Asia will, of itself, have a marked impact on greenhouse gas production.

Within New Zealand, transport will be hit hard as oil prices rapidly trend upwards. We have been talking about an enhanced public transport system for more years than most of us can remember but the fuel costs post peak-oil will make it much more financially attractive, and we will use wherever possible renewable energy from the sun and biofuels from waste products. Reduced access will mean that malls need to operate differently to survive and those next to an obvious main transport link, rail network and station will fare best. Bus routes will need to be linked to train stations and the buses will have purpose-built compartments for the storage and transportation of purchases from the malls, making it as easy as possible for shoppers to move their goods. Of course, many people will choose to buy through the Internet or whatever supersedes it, a trend that has been growing over the last couple of decades and that will accelerate rapidly after the oil shock. The conventionally-powered public transport system will, therefore, need to be supplemented with battery-operated delivery vehicles transporting goods for the last part of the journey, from the bus depot to the home.

Malls will need to morph into places that retain their meaning once the shopping binge is over. They will have to reduce their own consumptive footprint, especially their energy use because by 2020 all natural resources will have come under strong availability pressures, and the pollutants from consumptive processes will have been regulated. The Resource Management Act (RMA) of 1991 will have been fully implemented with clear national policy statements, standards and goals that leave businesses and communities in no doubt of their requirements. The RMA might also have been amended to further emphasise efficient resource use and to deal with the accumulative effects of contaminants. Paradoxically, despite cries of 'nanny state' and 'what about my property rights?' many people in business, including property developers, do not reject the certainty of well-defined rules and bottom lines. Of course, in 2010, there will be new issues to consider, such as the incorporation of low resource-use technologies and limits on what can be constructed, but with better definition of those limits, the time and money to get through the consenting process could be substantially reduced. Clear legislation could ensure that all businesses have the same regulatory advantage, the new rules creating parity even though they raise environmental expectations.

With higher requirements for resource and environmental management and fewer customers, malls will have to reinvent themselves as a different kind of community service. They will still have shops, but their primary focus on cheap imported clothes and other goods will have to change.

Hobson Mall in 2020
Future scenarios will always involve an element of guesswork; they tend to be a mix of what is aspired to and what is feasible, given what we know of future trends. One plausible future scenario could be that Hobson Mall – which we are imagining was built in 2007, takes the lead in reducing fossil fuel consumption and other activities that produce greenhouse gases. For the remainder of this chapter, we are going to imagine the world of Hobson Mall from 2020 looking backwards.

The mall owners had originally considered a hierarchy of options for an over-arching vision to set the scene and guide the design in 2007. The options were possible approaches to the environmental design of the mall and its operations. The further down the list, the greater the environmental benefits.

The options listed were:

1. 'Do only what is required. Do everything required to meet current code requirements.' This entailed unavoidable action, and carried no particular merit, as these actions had been codified by society as the minimum possible standards.
2. 'Go beyond legislation. Do everything that can be shown to be cost-effective in the short to medium term'. Any business that had planned for the medium term would have made all possible cost savings – the obvious 'low-hanging fruit'.
3. 'Show innovation and future thinking. Do things that will show

benefits overall, even if they may appear not to be cost-effective in a simple or single consideration.' This category represented the introduction of 'joined-up thinking', the intention to consider all the benefits of any action, not purely its financial payback.
4. 'Be a leader. Do projects that are highly visible, with measurable environmental and social benefits, even if their cost-effectiveness appears longer term than any of the above.' These are projects that make a development memorable and provide it with resilience against resource-based shocks and climate change.

When initially built, Hobson Mall, like most businesses in 2007, was still in one of the first two categories. Key performance indicators of success, which were focused solely on the financial bottom line, made it very difficult for managers and boards to sign off on innovative environmental features, which relatively untried at the time, were thought to carry greater than normal risk, even if intuitively the decisions seemed sound. This reluctance existed despite the fact that many organisations had a stated commitment to consider social and environmental impacts. More sophisticated financial systems and 'progress' indicators[3] were being researched to help business take the longer view of their role in society, but these had yet to be widely taken up.

However, within five years of its construction, the owners of Hobson Mall realised that they should have tried to aim further down the options list during the mall's construction; it would have cost them less to build sustainability into the mall than to retrofit it later. They acknowledged that things had changed much quicker than anticipated: that within a few short years oil had become very expensive and adaptation to and mitigation of climate change were starting to impact on every activity. In other words, they were faced with recognising that the International Panel on Climate Change (IPCC)[4] committee had been right.

By 2020, great strides had been made on the environmental front. No further malls had been built once the oil shock had hit in 2010. To survive, Hobson Mall decided to retrofit a wide range of environmental features. As mentioned, this was far more costly than having incorporated them during the design phase in 2007, but nevertheless necessary to limit operating costs and meet society's new and enhanced environmental expectations.

A staged programme was put in place to retrofit Hobson Mall. Gradually, other mall owners followed suit, creating a new social and environmental context for the form and function of malls.

Community consultation

The Hobson Mall owners brought the local community into discussions on the future of the mall. They wanted to know what would keep community members as customers. People liked the hustle and bustle of malls, just as they gained pleasure from the atmosphere of outdoor markets in previous generations; the community wanted to retain the atmosphere but focus more on community well-being. Members recommended that the mall become a centre for a wide range of civic engagement as well as traditional and Internet shopping.

As the owners embarked on the design process, a community consultation group, which included iwi, and members of local government and various representative community groups, was involved.

Design features that expressed community identity, culture, 'sense of place' and development (for example, locally significant motifs, artwork, flora and fauna) were incorporated, with local history being recounted and celebrated throughout the development. Education, especially about the natural environment, global and local social challenges, was integrated into Hobson Mall's activities.

In addition to enhanced public transport, safe linkages were made to adjacent communities, in particular through pedestrian and cycling connections to the surrounding residential areas. With transport being more restricted, there was a need for social spaces to bring people together. In particular, Hobson Mall helped meet that need, becoming a meeting place for farmers' markets. It also had local artisans working on site and provided a supportive environment for a number of new businesses, especially those built around green technologies such as sustainable energy, waste elimination and low emissions-producing goods. This arrangement was not new, with Curitiba in Brazil being a modern example.[5]

These developments moved the anonymous mall of 2007 to a vibrant, locally focused and supportive community-based amenity by 2020. Customers still visited the mall using local public transport, on foot or by bike but for a wider range of services with a lesser focus on goods. Given the new emphasis on community outcomes, the owners of Hobson Mall developed a joint-ownership model with local government to enhance the effectiveness of their delivery.

Building and resource use

Access to potable water was not generally considered an issue in 2007, although an awareness of its value as a resource was rising through impending water rate increases. The impacts of climate change and the link between potable water and the energy (and so greenhouse gas emissions) involved in cleaning water to the required standard meant that a serious effort to reduce the use of potable water at the mall had to be made.

Water-use targets were set by Hobson Mall's managers against which progress towards greater sustainability was measured. These surpassed the world's best practice. Greywater reuse in toilets and the gardens helped reduce the total amount of water required. Former concerns over health issues with these systems had been dealt with many years previously when Australia was forced to develop widespread wastewater recycling during a drought in the beginning of the century. Low-water-use appliances were selected and these, in conjunction with greywater reuse and the rainwater collected from the roof, meant that the mall easily met its total water demand. Water self-sufficiency became a strong feature of the mall.

A serious environmental concern for high-density construction such as malls is the amount of associated impervious surface and its impact on stormwater. Even with vastly reduced traffic in 2020, an unmodified Hobson Mall would still have contributed large quantities of poor quality water through a combination of bacteria, heavy metals and polyaromatic hydrocarbons (PAHs) into the piped stormwater network. These contaminants would have ended up in the estuary, leading to degradation of our coastlines. The owners decided on a vision of 'Only pure water leaves the site in minimal quantities'. New Zealand had been tinkering with low-impact stormwater management for many years. In fact, a raft of innovative designs had been available using a mix of roof gardens, rain gardens, detention tanks, other detention measures and permeable paving, but they were slow to catch on in mainstream construction. So, the mall was retrofitted with pervious paving to allow stormwater in the remaining car parks to be filtered through the soil. Residual run-off went through cuts in curbing into median strips, which were constructed as rain gardens, being planted with flax, reeds and grasses – which all absorbed the water and contaminants and added to the aesthetics of the environment. Some of the concrete roof became a roof garden where people could meet and relax.

Sewage was collected and treated in composting toilets that had generally been considered unsuitable for most urban areas in 2007. They were standard urban fittings by 2020 and provided an easy solution to managing human waste on-site. The proliferation of these and similar toilets was one clear indicator that people were prepared, once again, to take responsibility for their own environmental consequences. It was a shift in attitude that had started with recycling of household waste and then expanded, so that many of the services delivered remotely through major infrastructure in 2007 (energy, water, waste recycling), by 2020 were being managed on-site, at the household or community level. This did not mean a reduction in health standards; modern technologies meant that health was not at risk. There was a new approach to individuals' management of and responsibility for their environmental impact, one backed by technical solutions that remove risk and make good practice easy.

Minimising energy use was critical for the mall. The mall was retrofitted with light emitting diodes (LEDs), which use a fraction of the electricity of older types of lights. The mall was also fully insulated to a standard that was twice that of the 2007 Building Code.

The complex developed on-site renewable energy generation on a large scale. Not only did this provide enormous savings, but it was also a visible commitment to reducing carbon emissions, as well as a good educational and community resource.

Specifically, the mall had plenty of open space around it, which allowed the appropriate siting of a large wind turbine, which was often able to supply power, not only to the mall but to surrounding premises. With fewer car parks, the mall had ample roof space for locating solar panels to heat water and photovoltaics for making electricity in areas where there were no roof gardens. Solar water heating was standard practice by 2020. Also by that time, the cost-benefit ratios for many electricity-producing technologies including photovoltaics had improved for the better, and the technologies were more efficient. For example, by 2020, photovoltaic materials, especially through nano-technology, had become cheap and easy to retrofit so that most surfaces, walls as well as roofs, could absorb energy from the sun and turn it into electricity.

While there was a cost associated with retrofitting, once it was completed, and other energy-efficiency measures such as installing double glazed windows and high levels of insulation had been undertaken, the mall's operational costs were negligible. In fact, without these lower running costs, it would have been hard to keep Hobson Mall from financial ruin, given the required change in focus and cost structure of many of the goods and services it continued to supply.

No waste
One of the new functions Hobson Mall had taken on to survive, was to be a resource recovery centre within a 'green technologies cluster' where new uses were found for used products. The need to preserve precious non-renewable resources and make serious inroads into reducing greenhouse gas emissions had meant that many technologies emerged to reform materials that would once have been waste. All local building waste such as timber was recycled for other building projects or made into other products: waste plasterboard became a soil conditioner; concrete was restructured into pervious paving; glass was used as fabric. Much greater emphasis on recycling and reuse meant that there were the economies of scale to be made with the creation of these new industries locally. In addition, many products were now manufactured with reuse in mind, a cradle-to-cradle approach;[6] for example, no longer were electronic goods designed as sealed units which, once broken, had to be thrown away. Waste charging prices, introduced in 2008, eventually paved the way for financially viable resource recovery centres and for the design and manufacture of goods with re-usable components. Packaging, which had been the scourge of the early 21st century, was substantially less by 2020 and only packaging that could be recycled or reused was allowed by legislation.

Life-cycle planning
Of course, had the mall been planned with sustainable features in the first place, the owners would also have considered the life cycle of all the materials used in all phases of its construction and operation, including during:

- The design phase: the requirement to set performance targets. This would have been embraced and the mall designed to meet those targets.

- The operational phase: the performance and use of the materials over time. These would be measured on a regular basis, and compared regularly with the relevant targets.
- The end of life phase: the guarantee that no materials or components end up as 'waste', the goal being to leave the site available for another user.

Had the mall been built in 2020, it would have had to conform to a rating system that allowed it to assess how environmentally sustainable its design was, as well as monitor its operations against a set of key performance targets.[7]

THE FUTURE OF MALLS

So, has the mall a future in 2020 and beyond? The 2007 form of Hobson Mall reflected a lack of social and environmental awareness of its owners, but with sufficient retrofitting and refocusing it had become far more sustainable. Improved public transport meant that the mall was accessible without dependency on the private car and the mall had diversified to provide a much wider range of services, none of which were based on the provision of cheap goods. Hobson Mall also provided a community and educational focus and through the inclusion of more roof gardens and green spaces, helped bring biodiversity back into the city. With the advent of more renewable energy sources, the mall no longer required external energy sources and dealt with its own waste and minimised its waste load on others.

By 2020, malls will have reduced their carbon emissions to a sufficiently low level that only minimal off-setting of the remainder is required. Most of the environmental changes required are not radical; they could all have been incorporated in 2007. Those malls that remain successful, no doubt including Hobson Mall, will be those that take notice of their environmental and social responsibilities in the next few years. They will be the early adopters, who adapt to the future pressures of climate change and a 'carbon constrained world'.

ENDNOTES
1 Joint Working Group. 1999. *Saving Rates and Portfolio Allocation in New Zealand*. Treasury Working Paper 99/9. Wellington: New Zealand Treasury.
2 Hubert's graph and several other curves of predicted petroleum consumption can be viewed at http://www.hubbertpeak.com/curves.htm (accessed 8 Aug 2007).
3 Progress indicators are discussed at http://www.treasury.govt.nz/speeches/socialgoals/proxies.asp (accessed 21 Aug 2007).

4   To find out more about this committee, go to http://www.ipcc.ch/ (accessed 21 Aug 2007).
5   See Hayakawa, L., de Rocio Rosário, M. and Taniguchi, C. 2002. *Orienting Urban Planning to Sustainability in Curitiba, Brazil*. Toronto: ICLI, Local Governments for Sustainability at http://www3.iclei.org/localstrategies/summary/curitiba2.html (accessed 21 Aug 2007).
6   For more information on this approach and the book by W. McDonough and M. Braungart (2002, North Point Press), visit http://www.mcdonough.com/cradle_to_cradle.htm (accessed 21 Aug 2007).
7   One such rating system available in 2007 is 'The National Australian Built Environment Rating System' (NABERS), a system developed for the Australian Government as a monitoring tool for continual assessment of the performance of a building in operation co-designed by one of the authors of this chapter www.nabers.com.au (accessed 20 Aug 2007).

Computing away climate change

Alexei Drummond, John Hosking, Christof Lutteroth, Gerald Weber and Burkhard Wünsche

Alexei Drummond is a Senior Lecturer in Bioinformatics and his research focuses on computer models of evolution and ecology. These models can be used to better understand the relationships between climate, human activities and the genetic diversity of species.

John Hosking is a Professor of Applied Computer Science at the University of Auckland. His research focuses on software engineering and software tools, but he also has an ongoing interest in the social impact of computing. Particular interests in this area include Internet safety and technology transfer from academia to industry.

Christof Lutteroth has studied Computer Science at the Free University of Berlin, and is currently a researcher and lecturer at the University of Auckland. His research explores ways to make software development more efficient by applying methods and tools that are easier to use and understand.

Gerald Weber is a Lecturer in Software Engineering at the University of Auckland. His research is focused on modelling business logic and applications and on human–computer interactions. He is investigating how we can use information systems more efficiently, a question that is also important for the environment.

Burkhard Wünsche is a Senior Lecturer in the Department of Computer Science of the University of Auckland. He is the director of the Graphics Group and also directs the division for Biomedical Imaging and Visualization.

THE CURRENT SITUATION
Advances in computing and information technology are potent symbols of our rapidly increasing technological mastery of the world around us. With computers, we have launched space stations and satellites, and erected vast global information networks that connect people to each other and the Internet. Computers were used to assemble the first complete map of the human genome, and develop detailed models of global climate. These climate models have confirmed that greenhouse gas emissions – a by-product of human development – are driving rapid changes in the Earth's climate that may have far-reaching consequences for our future generations and many other species of life.

The information technology revolution is rapidly changing society. Looking to the future in an area of such rapid development is fraught with uncertainty. However, some trends are clear. Personal computing and information technology will continue to be driven by people's need to feel connected. Mobile devices like phones and music players will become ubiquitous and powerful, and computing will enable people to work with increasing flexibility. Computerised machines and automated processes will continue to replace humans to save time and money, increase efficiency and release workers from mundane and repetitive tasks.

THE VISION FOR 2020
By 2020, information technologies will allow us to routinely tele-work from home while still feeling 'as if we were there' with our co-workers. Aided by rich information that is available increasingly rapidly, New Zealanders will be able to easily make individual decisions that reduce their carbon footprints – and computers will enable developments in diverse areas such as monitoring emissions, on-line shopping, digital content, intelligent transport systems and scheduling.

STRATEGIES TO ACHIEVE THE VISION
Government, industry and popular support will be necessary for the development of computing advances and technologies that could play prominent roles in making New Zealand carbon neutral by 2020. Will tomorrow's computing deliver the power to change the world for the better? It could – but only if that is what we aim for.

WHERE WILL COMPUTERS TAKE US?

The year is 2020 and computers are ubiquitous and increasingly invisible – forming the backbone of an interactive and technology-based society. Exactly how the future will look is difficult to say, but what is undeniable is that the computer technology revolution will continue to be a major force for rapid changes in modern society.

The Internet, mobile phones, email, the iPod and many other high-tech innovations have profoundly changed the day-to-day lives of people in New Zealand and around the world. Computer technology has also transformed personal behaviour. We now rely on it for communication, navigation, banking, education, shopping, entertainment and our love lives. Internet dating is rapidly replacing traditional methods, as are social networking websites, which have been widely taken up by the current generation. In this new age of information, Wikipedia and Google can provide answers to many questions, answers that were previously inaccessible to the average member of the public. Most secondary school children have mobile phones and this trend is spreading into primary schools, as we become increasingly digitally connected to our friends, family and colleagues from an early age.

By 2020, everything from our toaster to our heart monitors will be connected by wireless networks and computers will aid us with almost all our daily tasks and decision-making. There will be significant changes to the ways in which we interact with information technology (IT) systems. The advent of cheap, large, flexible display technologies together with advancements in voice and gesture (for example, hand movement) recognition, and simulated touch (haptics) and smell mean that almost every room and room surface could act as an interface to available IT systems, and with very high degrees of realism and usability. Computers will indeed become part of the very fabric of our homes and buildings. This will significantly change the way we work, play, and interact with others. In the process, the technology will provide significant opportunities for reducing the carbon footprint of our everyday activities, primarily through reducing the need for travel, and its associated carbon costs, but also by allowing us to more rapidly monitor, understand and react to our carbon consumption and emissions in our organisations and daily lives.

Computers will also continue to be central to scientific research – foremost in the area of computer simulation and modelling. Today, using computer simulations, scientists and engineers model everything from the beating of a human heart to the action of earthquakes on skyscrapers. Computer models of space navigation were used to guide humankind to their first steps on the moon. Computer models of flight have aided the development of passenger jets that now fly people anywhere in the world in a day. Computer models of evolution have been used to trace back the history of the human species through our DNA to our origin in Africa.[1] In the field of climate science, computer models have been used to predict the effect of rising atmospheric levels of CO_2 and other greenhouse gases on global temperature and sea level.

TELE-WORKING

Commuting to work is a substantial contributor to the carbon footprint of many New Zealanders. Especially problematic is the increased use of air travel for business meetings in an increasingly global market. Jet aircraft burn vast amounts of fuel and business flights represent a rapidly growing component of the carbon footprint of many professionals. For example, based on the amount of fuel consumed per passenger we can calculate that a professional who flies once to Europe each year produces about 10 tonnes of CO_2. This amounts to half of New Zealand's average annual per person CO_2 equivalent production[2] and far exceeds the estimated maximum level for sustainability of annual per person CO_2 production (based on IPCC 550 parts per million Target).

Tele-working, or working from home and using IT resources to interact with work colleagues, is already a popular alternative to commuting. It has been estimated that as much as 20 per cent of workers in the USA make use of tele-working options.[3] By 2020, this will have significantly increased, as the availability of high-quality display technologies and high bandwidth network connections in homes will mean that working at home will provide essentially the same quality of social/business interaction as would occur when physically within the office. The high-quality teleconferencing capability that is currently available in very expensive commercial systems – such as HP's HALO system,[4] which uses very high bandwidth and cinematic techniques to create an 'as if you were there' experience – will be available within your home and will allow you to work, discuss business problems, and chat about the overnight sports results or celebrity gossip with your work colleagues just as if they were in the room with you. The resulting reduction in your carbon footprint primarily results from the elimination of the environmental costs of commuting, but also from reduced office space needs.

This means that fewer materials are used in constructing, furnishing and maintaining (for example, air conditioning and heating) extra office space. As tele-working increases, the relative savings in office energy costs will become comparable to the energy savings in transport.[5] And the time gained by no longer commuting for tele-workers, leading to increased business and/or social productivity. A downside is, of course, increased energy and water use within the home. For air-conditioned or heated homes, this will offset any equivalent gains from a reduction in office carbon needs. Nevertheless, for temperate climate countries, such as New Zealand, the carbon footprint reductions will significantly outweigh the increased carbon emissions from the home.

Similar tele-working methods will almost invariably affect the way in which education is delivered. Already universities and polytechnics worldwide are adopting more flexible delivery approaches, with significant on-line delivery of course material using both static resources, such as repositories of lecture notes and reference material, and more dynamic mechanisms such as on-line forums, and streaming or podcast-based delivery of lectures.[6] Institutions, such as the UK's Open University and the newer University of Phoenix (Arizona, USA), are arising that are almost purely on-line, with little if any campus-based teaching (typically the latter being limited to postgraduate research students). This trend will undoubtedly continue. Students will increasingly study from home, with the same 'as if you were there' motivation and resultant reduced carbon footprint, as will drive other tele-working behaviours. How much such behaviour extends to secondary and primary education remains to be seen. The environmental benefits are less obvious in this case, and the social costs likely to be higher.

The same and potentially increased benefits of tele-working extend to business travel other than to and from the place of work, be it across town to meet with clients, or out of town for business meetings. Increasingly, we will see such activity undertaken from either the office, as tele-conferencing, or tele-working from home. Events such as 9/11 have already significantly increased the use of tele-conferencing for out of town travel to eliminate both the time cost of travel and the perceived security risks, and have been major drivers for the development of high-quality teleconferencing systems such as HALO. The enormous environmental and health benefits of eliminating business-related air travel will become another important driver for this type of activity. While the number of people involved is far less than for standard tele-working, the impact is likely to be high due to the high environmental cost of air travel. Growing numbers of academic conferences are including an on-line interactive component or are being held exclusively on-line, leveraging high-speed academic networking systems, such as Internet-2,[7] or New Zealand's own KAREN[8] network. An alternative may be an increase in multi-node conferences, where multiple regional centres collectively host such conferences, allowing an element of direct personal interaction and social activities without the very high carbon emissions of a conventional international conference.

TELE-ENTERTAINMENT

In addition to work related activities, the new IT infrastructure will significantly affect our family and leisure activities. Virtual get-togethers with family and friends will become more common as the environmental cost of travel begins to be reflected in prices via carbon taxation and other governmental policy changes. Some social activities, such as sharing meals, are not suitable, but other aspects of socialisation will be able to be replicated using the same infrastructure as will support tele-working. Tele-entertainment, or obtaining entertainment resources on-line in the home, will be even more widespread than today. Movie theatres will become obsolete, with new features being released on-line rather than in theatre, and played in high-quality home theatres. Photo-realistic renderings of humanoid models and 3D environments will make it possible to develop virtual worlds for entertainment (for example, car racing games) to replace the corresponding real world activities. Environmental benefits for this type of replacement activity will be relatively minor in comparison to, say, tele-working, but still important. In the movie industry, computer graphics are already increasingly used to replace large film sets or expensive special effects. Augmented reality will become part of our everyday life. Already, many TV stations overlay broadcasted sports footage with virtual advertisements that eliminates the need for wasteful physical advertisement banners in stadiums.

ECO-FRIENDLY TELE-TOURISM

Tourism, as a sector, will be heavily impacted by environmental concerns. The carbon-related costs of tourism are heavily dominated by air travel and, due to its remoteness, New Zealand will be seen to engender a much higher carbon footprint to visit (as will, of course, travel by

New Zealanders elsewhere).[9,10] This is likely to be highly detrimental to our tourism industry and the New Zealand economy. More and more, tele-tourism or virtual tourism products will compete with real tourism owing to their low relative environmental cost.[11] Virtual tourism products will be of high quality and will be highly interactive, providing a far more competitive alternative to the 'real thing' than current tourism documentaries. This offers an opportunity for New Zealand to capitalise on its creative industry's expertise, to construct such products based on our country's natural beauty.

However, virtual tourism products are highly unlikely to replace real tourism in a significant way. The tendency instead will be for tourism to be more regionally oriented, with local tourism opportunities being promoted on the basis of their low environmental impact. For New Zealand to compete, tourism operators will need to fully understand the carbon-related costs of their activities and seek to mitigate them as effectively as they can. IT can assist in this. If there are still significant regional costs, in the form of activities such as scenic flights, these could be offset at the site by, for example, virtual alternatives. Eco-friendly accommodation, using IT systems to monitor and optimise energy consumption, and carbon sequestration activities, such as tree planting, are other alternatives.

There are also alternatives to air travel. For example, plans have been proposed for hydrogen-powered container ships capable of travelling at 120 kilometres per hour carrying 600 containers.[12] Such plans could be adapted to suit passenger travel instead, with low carbon-cost fleets of high-speed passenger liners increasingly replacing long distance aircraft. The downside, of course, is that travelling as far as New Zealand will take a significant amount of time. IT solutions can help make this a viable choice for overseas tourists. Such tourists will come less often but stay for longer, and split their time between tourism and tele-working, using facilities wherever they are at a particular time, to maintain their working life at a part-time level. The line will thus blur between tourism and work residency, requiring more flexible visa rules for a country like New Zealand to accommodate the needs of this type of tourist. The accommodation industry will likewise need to adapt, to meet the need for larger amounts of suite-style accommodation, complete with high-quality IT infrastructure. Without such creative approaches, New Zealand's economy will be significantly and adversely affected.

PREDICTING THE WEATHER IN 2020

Climate models simulate the interaction between the oceans, land surface, ice, atmosphere and the sun. This is achieved by using complex mathematical equations that describe physical processes in the Earth's atmosphere and their dependence on human-made and natural parameters.[13]

Natural drivers of climate change include solar activity and atmospheric aerosols released by forest fires and volcanic eruptions. Human drivers of climate change include increased emission of greenhouse gases such as CO_2 from fossil fuel usage and changes in land use, and methane and nitrous oxide emissions primarily caused by agriculture. The major drawbacks of current climate simulations are the large uncertainties in their models. These are caused by: insufficient input data (lack of historical data and sparsely distributed measurements); inadequate knowledge of the physical processes governing the interaction of parameters that cause climate change; and the coarse-grained space and time resolution of the simulations. This means that experts differ on the predicted extent of climate change. This uncertainty has allowed sceptics to cause confusion in the public domain about the state of the science. However, while there are still uncertainties in the rate of future climate change, there is an extraordinary level of certainty and consensus about the direction of climate change and its immediate consequences.

In the future, more advanced climate models, increased knowledge of climate-influencing processes, more detailed input data and considerably more powerful computers will allow scientists to develop more reliable and comprehensive climate models. Successful forecasting of short-term weather and climate changes will decrease the effectiveness of sceptical attack, which in turn will make it easier for governments to introduce widely supported environmental policies. Advanced climate models will be especially important for New Zealand, since tourism, agriculture and forestry (all very climate sensitive) are currently our three largest export earners.[14] Detailed long-term weather forecasting and improved agricultural management techniques will make it possible to select suitable crops which minimise the need for irrigation and the use of pesticides. Early detection of El Niño and other destructive weather patterns will allow farmers and the seafood industry to plan harvesting activities and equipment, and staff requirements. The urban population will also benefit from improved climate models and weather forecasting. Smart homes will utilise weather predictions to determine heating and ventilation needs and coordinate the use of solar panels in order to minimise electricity consumption.

SMARTER DESIGNS TO REDUCE EMISSIONS
Mathematical models and the simulation of real-life processes and events are now an integral part of modern life. Computer Aided Design (CAD) and Finite Element Modelling (FEM) techniques are used to develop and test goods on the computer before production. This makes it possible to design, for example, cars that are more energy efficient and wind turbines that are more effective. Computer simulations are also used to develop improved production processes, so that they require less water and energy and create less waste.

In the future, these simulations will become more advanced and enable the development of new technologies. In the field of nanotechnology, which involves designing molecule-sized machinery and active compounds, computer simulations will be used to design cheaper solar panels, more effective hydrogen fuel cells and batteries, lighter and stronger materials, more effective filters and catalytic converters and better insulation materials. All of these advances cut energy consumption and the emission of greenhouse gases.

New computational models in bioinformatics will help us to better understand the structures of genes (genomics) and proteins (proteomics), which are fundamental to attempts to develop insect- and climate-resistant plants and plants with higher crop yields. The resulting reduction in agricultural land and transport requirements will help to curb carbon emissions. In the future, this research will advance dramatically and we will see the engineering of plants for producing higher quality biofuels and the development of plants for reducing soil salination and soil decontamination.

Cheminformatics, the use of information technology in the field of chemistry, will continue to increase in importance since there are more than 10^{60} (10 with 59 zeros after it) possible molecules[15] and computers will be necessary to identify or engineer molecules that are effective against new or resistant diseases that spread more rapidly due to climate change.

Operations research uses mathematical modelling, statistics and algorithms to optimise the coordination and execution of the operations within an organisation. Its algorithms are already commonly used for scheduling problems, for example, when staffing airplanes. In the future, operations research will become indispensable with everything from parcel delivery, traffic flow, vehicle utilisation, urban planning and the heating and ventilation of buildings. It will be used to optimise these processes in order to minimise energy usage.

MONITORING EMISSION AND PROVIDING FEEDBACK
In the world of engineering, it is an acknowledged fact that in order to control something, we first need a way to measure it. This applies not only to engineering, but also to the environmental impacts of our everyday lives. If we know how much energy we consume and the amount of CO_2 emissions we are responsible for generating, we can compare these measures against national and international statistics. Comparatively high measures mean that there is still room for us to reduce our CO_2 emissions (and often to also save time and money). Monitoring our emissions enables us to evaluate the changes we make. After a change, we can objectively decide whether the change affected CO_2 emissions in a positive or negative way. It also means that success can be made widely known and positive change can be rewarded with public accolade or governmental support.

In Europe, plans are underway to replace all electricity and gas meters with so-called 'smart meters'. A smart meter allows consumers and energy producers to monitor energy consumption remotely in real time. Consumers can check at any time how much energy they currently consume, how much they have to pay for it and how much CO_2 it produces. Energy producers can use this information for exact billing, without having to send in service personnel to read each meter manually. This in turn saves on personnel costs, so that energy providers can recoup the initial investments necessary to replace old meters by smart meters, and operate more efficiently in the long run. According to the UK consumer body Energywatch, smart meters help consumers to reduce their costs as well. In Italy, where smart meters have been installed since 1997, consumers were able to cut their power demands during peak times by 5 per cent.[16] Recently, it was announced that from 2008 onwards households in the UK will be able to request a free device for real-time electricity monitoring in lieu of the widespread adoption of smart meters. These devices consist of a sensor that can be easily installed on old power meters, and a portable display that shows data such as the current power consumption, cost and CO_2 emission at any time. Such schemes are now being piloted in New Zealand as well,[17] with the expectation of a similar positive impact on CO_2 emissions and cost savings for energy producers and consumers equally.

Besides monitoring energy consumption, such a scheme could be backed up by computer-supported services that help people analyse and improve their energy consumption. Instead of a smart meter just delivering real-time data to a consumer, it could also identify energy wasters in the household, such as overly power-hungry fridges and other inefficient stand-by devices. Software could use statistical data to tell consumers whether a device consumes more power than other devices of the same kind, and how much such a device should usually consume. Furthermore, it could recommend energy saving measures such as using timer-controlled lights, and calculate how much energy and money would be saved, and CO_2 emissions reduced, by implementing them.

Exact power monitoring with smart meters could also be utilised to implement powerful regulation strategies. With accurate metering, taxation schemes could be introduced to target individuals and organisations that cause disproportionately high CO_2 emissions, and reward those who manage their emissions with care. National energy consumption could be made more transparent by publishing statistical data on the Internet for short time intervals, and inspire people to put more effort into saving power. For example, the 2007 'Hamilton energy blitz' campaign published the region's electricity consumption daily in the Waikato Times, giving immediate feedback about the campaign's success.[18] If data about the whole country were made available, they could give rise to friendly competition between different regions. People would try not to waste energy, and success could be publicly acknowledged. With a permanent power-monitoring infrastructure in place, such competitions could become a permanent strategy for promoting and rewarding energy-conscious behaviour.

SAYING IT WITH PICTURES

Technological advances over the past decade have enabled scientists to create ever larger and more complex scientific datasets. Examples are advanced numerical simulations, satellite measurements and medical imaging data. As a result it has become increasingly difficult to understand, analyse and communicate the resulting information. Scientific visualisation is an attempt to achieve this goal by transforming numeric scientific data into one or more images that, when presented to a human observer, convey insight or understanding of the data. Visualisations are already

regularly used in science and engineering to analyse large amounts of measured and simulated data. They are increasingly used by non-scientists to make policy decisions and for educating laypeople about scientific discoveries such as the effects of climate change.

In the future, displays will become ubiquitous, the information will be richer and advanced visualisation techniques will make it possible to compare multiple data sets and display uncertainty, which makes correct interpretation of the data more likely.[19] This, in turn, will enable decision-makers in politics and business to develop policies and business plans to reduce climate change and to implement them efficiently and effectively. Visualisation and simulation will be interconnected (computational steering) in order to speed up product development cycles and so that government, businesses and we can react faster to predicted events. As a result, new environmentally friendly products and procedures will be developed which conform to new policies. Examples are: lower emission cars; more effective and cheaper micro-generators and solar panels; pricing models for water and energy that encourage savings but do not disadvantage people on low incomes; and superior integration of transport systems for large cities.

Finally, computer graphic techniques will also help in the analysis of imaging data obtained from video surveillance systems and satellites. Image analysis and feature-tracking technologies will advance dramatically and, in combination with machine learning algorithms, will make it possible to automatically detect illegal logging, forest fires, changes in glacier size and Arctic ice cover, and polluters (for example, oil tankers illegally cleaning their tanks).

ELECTRONIC CONTENT, IN THE PAPERLESS OFFICE AND HOME

Downloading content and reading it electronically naturally reduces resource consumption owing to the elimination of environmentally costly paper. For example, reading news on-line has about 50 times less environmental impact than reading a physical newspaper,[20] even allowing for the environmental impacts of producing the electronic device. So, widespread replacement of paper by electronic resources is an obvious step toward carbon neutrality. Already, we see a significant switch away from paper in the area of photography. The advent of digital cameras has meant that many people now rely on electronic rather than physical photo albums, eliminating the paper and chemical costs of conventional photo processing.

It is not easy to switch to the paperless office, however; many of the problems of doing so are quite challenging to computer science researchers, in particular to specialists in human–computer interactions. Currently, on-line displays lack the fine-grained resolution, portability and flexibility of paper. However, the new high-resolution, flexible display technologies currently under development will, by 2020, eliminate these disadvantages, permitting, for the first time, 'electronic paper' to compete effectively with conventional paper.

Computer science can contribute further by creating more intuitive user interfaces for the handling of all electronic content. We must move beyond the simple file system, and search functionality must be a fundamental part of the change. We do not yet have convincingly effective formats for archiving data, something we urgently need for the long-term storage of digital data. Nor do we yet trust electronic signatures or authorisations of important documents, meaning we still need a paper trail in our office work. Current business behaviour does not help; for example, if you accept your bank's offer so that you stop receiving paper statements, you either have to print them every month, or you lose information after three months. But the situation is changing, and one can be confident, that in 2020, we will have achieved a mostly paperless office and a mostly paperless home.

One of the advantages of the paperless office is that it can make our office work more efficient and flexible. Computer scientists can help in getting computer technology to deliver on its promises: that we will find our documents easier; that we will be able to share them intuitively; that we will be able to work on them collaboratively. The paperless office will make it easier to move the office: we will be able to work at home and while we are travelling.

ON-LINE SHOPPING

Conventional shopping is an activity rich in sources of carbon emissions. For example, for Amazon.com's delivery of a newly released Harry Potter novel, FedEx proudly declared they needed 100 planes and 9000 trucks.[21] Overnight shipping is a particular matter of concern, since it often involves shipping by air. How will we do our shopping in 2020?

The question of whether on-line retail is ecologically better or worse than conventional brick-and-mortar retail can be evaluated reasonably well by creating eco-balance sheets of all the material streams involved (see below). But studies have shown that relative carbon cost-benefits are crucially dependent on optimisations in the respective business model and in transport. Depending on slight variations, either alternative can be environmentally better. Looking again at books, for both business models, the fuel costs for delivery outweigh the production cost by a factor of more than four.

Which model is more efficient depends on the impact of personal transport on the overall carbon balance. If the customer makes a 10-kilometre return trip just for a single book, the on-line retailer would be the better choice. If, however, the transportation for buying the book is negligible, then the brick-and-mortar retailer is better.[22]

Future developments such as the inexorable rise in fuel prices will increase the costs associated with both models, creating a real chance for electronic content to replace traditional products where feasible. In the future, your bookshelf will be replaced by a flexible, tactile 'bookplayer' in much the same way as an MP3 player has already replaced your CD rack. Books will be downloaded over wireless networks onto the bookplayer as and when you need them. For other types of product, the sorts of optimisation techniques described earlier and the intelligent transportation options discussed later will be important tools in minimising the carbon emissions of product distribution.

New, smart, low carbon-emitting products

One of the things that computer science can contribute to reducing the carbon emissions associated with conventional products is to effectively track the carbon balance of products and inform the customer about them. By 2020, customers will be able to learn about the carbon impact of a product from their shopping cart both in terms of the item's production and its lifetime. This system will be integrated with the solutions described in the section on Monitoring. In particular, electronic products are dangerous waste – in the future, they will be tracked over their lifetime. Although trade from Internet auctions like TradeMe already accounts for a considerable amount of New Zealand package delivery,[23] such second-hand solutions will become even more attractive, for buyer and seller, wherever it helps to reuse products. Also, by 2020, it is quite likely that the three-year-old washing machine you are going to buy on-line will tell you whether it was working in the last few months, and whether it was really used only once a week or rather did the laundry for the whole neighbourhood. You will not have to physically travel to inspect it; a tamper-proof status report will be on the auction site and you, the prospective owner, will be able to decide based on the data provided. As another example, you will not ask the previous owner about the fuel efficiency of a car: instead you will ask the car. It will be able to tell you precisely, once it has switched into a for-sale mode.

At the moment, a huge amount of packaging ends up in our landfills. In the future, in-package sensors will show you whether a package was handled with care by the courier service. Delivery packages will be able to use less styrofoam because the packages will be handled more carefully. If you go shopping, why not pick up the parcel or the grocery for your neighbour? You will receive an attractive electronically-transferred payment for your trouble as soon as the parcel picks up your neighbour's door's Radio Frequency Identification (RFID) signal. If a thief takes a chance, the alarm will blare as soon as the parcel's connection to the receiver RFID fails. It is the same technology that will protect parked bicycles and will make it unnecessary to crawl behind the bike and secure a lock next to the greasy chain. E-bikes will be a very attractive vehicle in the future. There will be even be rental E-bikes that you will be able to leave wherever you want and that you can pick up wherever you see one.[24]

The concept of smart products can also be extended back through the supply line to track the primary resources used. This will ensure that the product you buy has been produced sustainably. For example, to avoid consumer complicity in illegal logging and the resultant wider environmental problems,[25] in the future, every logged tree could be in a database, and have a genetic fingerprint. When its timber turns up in an upmarket furniture shop, consumers will be able to tell exactly where it has come from.[26]

Computer science will also help us manage many of our future everyday consumables. Why not introduce the 'NZ Mug', a reusable stoneware mug used by all coffee outlets for takeaways that can be returned everywhere.[27] It would put an end to cardboard disposable mugs. A fee could be raised to guarantee return, but it would be repaid electronically if you put the mug back onto the return tray.

INTELLIGENT TRANSPORTATION SYSTEMS

As transport is such a large contributor to our total carbon-emissions output, many smart products, optimisations and changes need to be made in this sector. The most direct strategy is certainly transport avoidance, and this topic is addressed in a number of sections in this chapter and in previous chapters. However, there will still be a considerable demand for transportation in 2020, and this needs to be dealt with in an environmentally responsible manner.

A person's contribution to transport-based carbon emissions can vary widely, depending on the types of transport used, how much they are used and when. For example, to a person using a private car, it makes a big difference whether the car is used during rush hours or not. During rush hours, a trip can potentially take several times longer than during off-peak hours owing to traffic congestion. Traffic monitoring can help people to avoid congested traffic and reduce the time they spend on the road. In the Auckland region, where congested traffic is very common, infrastructure for traffic monitoring is already in place. The Transit New Zealand website[28] shows the current traffic situation on motorways 1 and 16, where traffic congestion usually occurs. However, it does not show historical or aggregated data, such as the average congestion at a particular time of day.

On-line traffic monitoring services could be extended to help people find the most suitable times, routes and methods of transportation. So, statistical data about the traffic congestion on an average working day could help people to choose better times to travel. Such a system could go as far as asking users about their time constraints and then recommending good alternatives, including public transport. Simple traffic prediction and cost-estimation methods could be applied to show travel duration, cost and CO_2 emissions for each alternative. Since cars keep on burning fuel even in congested traffic, avoiding congested traffic not only saves time and money, but also reduces CO_2 emissions and is less unhealthy. In the future, more cars will have onboard computers with global positioning systems, and it is very likely that by 2020 traffic data will be broadcast in digital form. As a result, cars will be able to show real-time data about emissions, cost and traffic to their drivers, and use that data to guide them around areas of traffic congestion.

Another step towards making people more aware of their CO_2 emissions is to give them more feedback about their cars. This can be achieved with the help of a computer-supported information service such as a web portal. Land Transport New Zealand already offers such a portal,[29] and many people use it, for example, to renew their licence on-line. The system could make additional information available from a database, such as the efficiency of different car models according to their emissions. Since car efficiency can be described in terms of running costs, this would not only raise awareness about emissions, but also help people to choose cars that save them money in the long run. Making statistics publicly available can help to motivate people to reduce their emissions. Such statistics could include the numbers of low-emission cars and high-emission cars per region, or the amount of CO_2 released each month. Similar to electricity consumption, this could possibly inspire a positive spirit of competition between different regions.

'Intelligent transportation system' is a general term encompassing new developments that make individual or public transport more automatic and convenient. Other terms are 'vehicle infrastructure integration', 'vehicle platooning', and 'smart roads'. A pilot project of this type was the California PATH project.[30] In the most straightforward of these types of system, cars switch into automatic mode in dedicated lanes (for example, on highways). In these lanes, the cars group into tight daisy chains, called 'platoons', which operate like a single vehicle. The tight sequence reduces drag and thus saves energy, and since following distances can be reduced, it maximises the number of vehicles on the road.

The platooning happens by either mechanical coupling or electronic distance control. Other positive effects of such technologies are a possible reduction in traffic accidents – around half of the traffic accidents happen on fast roads amenable to this technology – and increased lifespan of the vehicles through better operation. By automating cargo transport, another major road hazard can be reduced. Finally, once the smart road technology is introduced, an evolution towards a different, less harmful road construction might follow. Why? Smart roads are ideal places to implement an inductive power supply – the car obtains its power from a wireless electricity supply while it is in motion. This technology has been intensively studied at the University of Auckland.[31]

This technology is possible today thanks to standard computer technology that is already very commonplace. The remaining difficulty is that its introduction requires substantial infrastructure and legal changes, and concerted – perhaps even global – action. On roads, the technology can be retrofitted. On current cars, this might be difficult, but if this trend was actively anticipated now, new cars could be prepared for retrofitting. Computer scientists can help in plotting good standards and making the central software components open, so that patent issues do not hamper adoption.

Nevertheless, platooning technology also bears carbon risks, since it would make possible faster vehicle speeds. This potential might be attractive for consumers, who might create pressure to legalise faster platoon speeds. However, ensuring that the relevant legislation is designed with energy efficiency as its primary consideration would short-circuit this problem.

There are, of course, a number of reasons why such a technology may turn out to be unpopular. People might fear a loss of autonomy. Such a system lends itself naturally to tolling, creating cost concerns. Rush hour platoon slots may end up being auctioned off. Nevertheless, the carbon-associated benefits of such systems are compelling.

CONCLUSION

Computers and computing technology are powerful tools that have the capacity to greatly assist in the creation of a future New Zealand that is truly clean and green. But their assistance is not assured and does not come for free. Computers, mobile phones and other high-tech gadgets are also potent symbols of the modern penchant for disposable commodities and profligate waste, with many personal computers and mobile phones being discarded after one or two years of use. Like many high-tech manufactured goods, computers are extremely energy intensive in their construction, and thus, for their size, contribute disproportionately to greenhouse gas emissions. Today, the construction of an average computer uses around 260 kilograms of fossil fuels, 22 kilograms of chemicals and 1500 kilograms of water, for a total of 1.8 tonnes of materials.[32,33] That makes the ratio of fossil fuel to product weight 11, or an order of magnitude higher than the factor of one to two for many other manufactured goods such as cars and refrigerators. For these reasons, in the future, computer technology can just as easily contribute to the problem as it can to the solution. If we are going to succeed in the vision of computing away climate change, we first need to ensure that we have the policy and consumer behaviour in place to ensure that our computing devices are manufactured efficiently, and are used and reused much more effectively than they currently are, with sustainability and energy efficiency as our guiding principles.

While there are plenty of reasons to believe that computer science could be a significant part of the solution to our ambitions of carbon neutrality by the year 2020, there is no guarantee. Many of the potential solutions outlined in the chapter are not foregone conclusions and require support at government, commercial and consumer levels. While there is a great possibility for New Zealand to embrace sustainable technologies that will assist in computing away climate change, there are significant challenges to realising this ideal. We are reliant on the long-term vision and policy decisions of overseas companies and governments for the development of many of the industrial technologies required. Our vision will also require government and commercial support to continue to develop capacity in the IT and knowledge economy of New Zealand if we are to make the solutions in this chapter a reality by 2020. If used wisely, computer science is a fantastic tool that can greatly assist us in achieving our ambitions for a sustainable future.

ENDNOTES

1   Cann, R.L., Stoneking, M., and Wilson, A.C. 1987. Mitochondrial DNA and human evolution. *Nature 325*, 31–36. Available at http://dx.doi.org/10.1038/325031a0 (accessed 26 Sept 2007).
2   The figure is given as 19.4 metric tonnes CO_2 equivalent per person. UN Human Development Report, 2003. http://globalis.gvu.unu.edu/
3   Irwin, F. 2004. Gaining The Air Quality And Climate Benefit From Telework, http://safeclimate.net/business/solutions/teleworkguide.pdf (accessed June 11 2007).
4   Video Collaboration, HP Halo Collaboration Studio, http://www.hp.com/halo/index.html (accessed June 11 2007).
5   Matthews, H.S. and Williams, E. 2005. Telework adoption and energy use in building and transport sectors in the United States and Japan. *Journal of Infrastructure Systems March 2005*, 21–30.
6   Educause provides very good resources for studying the impact of technology on higher education: http://www.educause.edu/ (accessed 26 Sept 2007).
7   For more information, visit http://www.internet2.edu/ (accessed 26 Sept 2007).
8   For more information, visit http://www.karen.net.nz/home/ (accessed 26 Sept 2007).
9   Hunter, C. and Shaw, J. 2005. The ecological footprint as a key indicator of sustainable tourism. *Tourism Management 28* (2007), 46–57.
10 Becken, S. and Patterson, M. 2006. Measuring national carbon dioxide emissions from tourism as a key step towards achieving sustainable tourism. *Journal of Sustainable Tourism 14* (4), 323–338.
11 Cheong R. 1995. The virtual threat to travel and tourism. *Tourism Management 16* (6), 417–422.
12 Veldhuisa I.J.S., Richardson, R.N., and Stone, H.B.J. 2006. A hydrogen fuelled fast marine transportation system. Paper presented at the World Hydrogen Energy Conference 16, held on 13–16 June 2006, Lyon France.
13 Intergovernmental Panel on Climate Change. 2007. The Physical Basis of Climate Change. IPCC Working Group 1 AR4 Report, 22 May 2007. See http://ipcc-wg1.ucar.edu/wg1/wg1-report.html (accessed 25 Sept 2007).
14 Statistics New Zealand. June 2007. New Zealand External Trade Statistics. See http://www.stats.govt.nz/products-and-services/ext-trade-stats/default.htm (accessed 25 Sept 2007).

15 Kirkpatrick, P. and Ellis, C. 2004. Chemical space. *Nature Reviews Drug Discovery* *432*, 823–865. See http://www.nature.com/nature/insights/7019.html (accessed 25 Sept 2007).
16 http://news.bbc.co.uk/2/hi/business/4613613.stm
17 http://www.meridianenergy.co.nz/AboutUs/News/Central+Hawkes+Bay+households+first+in+New+zealand+to+get+smart+electricity+meters.htm
18 http://www.envirocentre.org.nz/index.asp?pageID=2145848835
19 Johnson, C. 2004. Top Scientific Visualization Research Problems. *IEEE Computer Graphics & Applications 24* (4), 13–17.
20 Toffel, M.W. and Horvath, A. 2004. Environmental implications of wireless technologies: news delivery and business travel. *Environmental Science & Technology ACS, 38(11)*, 2961–2970. See http://dx.doi.org/10.1021/es035035o (accessed 25 Sept 2007).
21 http://www.fedex.com/us/about/express/pressreleases/pressrelease071000a.html?link=4 (accessed 25 Sept 2007).
22 Matthews, H.S., Hendrickson, C.T. and Soh, D. 2001. The net effect: environmental implications of e-commerce and logistics. *In Proceedings of the 2001 IEEE International Symposium on Electronics and the Environment 2001*, 191–195. See: http://ieeexplore.ieee.org/xpl/freeabs_all.jsp?isnumber=19985&arnumber=924525&count=55&index=33 (accessed 25 Sept 2007).
23 Fox, A. 2005. TradeMe clients bolster NZ Post parcel volumes. *The New Zealand Herald*, Monday, August 29 2005. http://www.nzherald.co.nz/organisation/story.cfm?o_id=274&ObjectID=10342863 (accessed 25 Sept 2007).
24 DeMaio, P., Gifford, J. Will Smart Bikes Succeed as Public Transportation in the United States? Journal of Public Transportation, Vol. 7, No. 2, 2004.
25 anon. 2007. Orangutans 'face greater threat'. Available at http://news.bbc.co.uk/2/hi/science/nature/6337107.stm (accessed 25 Sept 2007).
26 Rhodes, W.M., Allen, E.P. and Callahan, M. 2006. *Illegal Logging: A Market-Based Analysis of Trafficking in Illegal Timber*. Final report for the U.S. Department of Justice Document No.: 215344, ASP TR-002, August 2006.
27 A prototype of a similar mug re-use system can be seen at HP's Cooltown centre in Singapore.
28 The address for Transit New Zealand's website is http://traffic.transit.govt.nz/. (accessed 25 Sept 2007).
29 Land Transport New Zealand's web portal is http://www.ltsa.govt.nz/ (accessed 25 Sept 2007).
30 Shladover, S.E., Desoer, C.A., Hedrick, J.K., Tomizuka, M., Walrand, J., Zhang, W.-B., McMahon, D.H., Peng, H., Sheikholeslam, S. and McKeown, N. Automated vehicle control developments in the PATH program. *Vehicular Technology, IEEE Transactions* 40 (1), 114–130. See: http://ieeexplore.ieee.org/iel1/25/2453/00069979.pdf?isnumber=2453?=STD&arnumber=69979&arnumber=69979&arSt=114&ared=130&arAuthor=Shladover%2C+S.E.%3B+Desoer%2C+C.A.%3B+Hedrick%2C+J.K.%3B+Tomizuka%2C+M.%3B+Walrand%2C+J.%3B+Zhang%2C+W.-B.%3B+McMahon%2C+D.H.%3B+Peng%2C+H.%3B+Sheikholeslam%2C+S.%3B+McKeown%2C+N (accessed 25 Sept 2007).
31 Auckland UniServices. 27 June 2003. Auckland research wins major Australasian award. See http://www.uniservices.co.nz/pageloader.aspx?page=812d3d0d76 (accessed 25 Sept 2007).
32 Williams, E. 2004. Energy intensity of computer manufacturing: hybrid analysis combining process and economic input-output methods. *Environmental Science & Technology 38*(22), 6166–6174. See http://dx.doi.org/10.1021/es035152j (accessed 25 Sept 2007).
33 Hirsch, T. 2004. Computers must be greener. Available at http://news.bbc.co.uk/2/hi/technology/3541623.stm (accessed 25 Sept 2007).

Deep organics

Brendan J. Hoare and Keith Thomas

Brendan Hoare has 24 years' experience in sustainable organic systems practice, design, leadership, theory and philosophy. Six of these were spent exploring traditional food production systems in Asia. Apart from his commitments to the Pacific Centre of Sustainable Communities, School of Natural Sciences, Unitec New Zealand (Auckland), he is currently a World Board Member of the International Federation of Organic Agriculture Movements, a member of Unilever's Sustainable Agriculture Advisor Board, Founding Director of the new Journal of Organic Systems and works in partnership with international and national projects. His passions lie in facilitating development of cooperative actions that will enable the worldwide adoption of organic systems, developing family land using perennial polyculture systems and surfing waves in remote locations.

Keith grew up on a typical suburban section in Papakura in the 1960s, which was self-sufficient in vegetables. This year he graduated from Unitec with a Certificate in Horticulture and a renewed commitment to sustainable food production. He has a large vegetable garden and is currently helping other people grow their own organic food.

THE CURRENT SITUATION
Industrial farms use monocropping, external fertilisers and pesticides to maximise food production. As a result, many of our first-class soils have become eroded and lost much of their original fertility. We are also failing to utilise the capacity of healthy soil, rich in microorganisms, to hold the carbon currently contributing to global warming. In our towns and cities, food scraps go to landfills, where they produce methane, rather than being fed back into food production systems. Food travels long distances from the farm to the table, further adding to the carbon debt created by our current agricultural practices. There are, however, pockets of people throughout Aotearoa/New Zealand who are rethinking current practices and showing the way towards a more sustainable future.

THE VISION FOR 2020
By 2020, our agricultural practices will have shifted significantly towards organic principles that encourage soil health, ecological diversity and self-sustaining systems. Our urban areas will be reclaimed as vibrant growing systems, with individuals, communities and local authorities planting a wide variety of fruit trees, food crops and carbon-dense native hardwoods. Children will get hands-on experience at growing and learning about eco-systems, soil and natural cycles. Aotearoa/New Zealand will be showing the way in a growing international movement towards organics and beyond.

STRATEGIES TO ACHIEVE THE VISION
To achieve this vision we need examples of these practices available to everyone who wants to get involved. Widespread education is essential, so that people learn the basic principles that go into maintaining healthy soil, sustainable land use and organic gardens. We need to do away with industrial farms and instead put in place systems that work with nature and approach problems as disturbances to feedback loops. Local government needs to provide the infrastructure and incentives to encourage optimal use of the land and recycling of waste. Most of all, we need to realise that as a nation we must work collectively to create this future.

INTRODUCTION

'To busy oneself with what is futile when one can do something useful, to attend to what is simple when one has the mettle to attempt what is difficult, is to strip talent of its dignity.'[1]

Jose Marti

Many civilisations have successfully worked the land for thousands of years, feeding large populations without factory-produced external inputs such as fertilisers and pesticides. In these ancient cultures, such as those of China, Korea, Japan and Taiwan,[2] growing was based on the principle of constantly enriching the soil, underpinned by the simple maxim of 'healthy soil, healthy food and healthy people'. Food production was not constrained to rural areas largely bereft of people, but was part of the mix of human activity. Even today, if you travel across much of Asia, you will see things growing in almost every available space – a small patch of vegetables on the side of the road, chickens foraging among a cluster of houses, fruit trees along the banks of a stream. In these growing systems, food and waste production literally feed off each other, as food scraps are composted and put back in the soil, and even human waste is often utilised as a rich source of manure.

Many of today's problems are a result of this cycle being broken. Most food production is reliant on non-sustainable, generally oil-based inputs. Food scraps largely go into landfills where they produce methane, a powerful greenhouse gas. Human waste, now contaminated with the variety of toxic chemicals we use to treat our bodies and clean our homes, is taken to sewerage plants, processed and disposed of far from sites of food production.[3] In addition, we are seldom engaged with the growing of food, finding ourselves disconnected from how or where it is grown.

One consequence of this shift in the cycle of food production, consumption and waste treatment, is a disruption to the carbon cycle. It has been estimated, in fact, that half the current excess (above pre-industrial levels) of CO_2 in the atmosphere, is due to the combination of soil degradation, modern agricultural practices and food miles (a measure of the distance and methods by which food travels from farms to the people that eat it).[4] There is huge potential to reverse much of this in Aotearoa/New Zealand, a potential we must tap if we are to become carbon neutral.

THE SOIL: A CARBON SINK AND THE KEY TO FERTILITY

Carbon dioxide is a combination of carbon and oxygen. When carbon-intensive life forms are encouraged, the carbon is removed from the air, leaving the oxygen. The process of turning airborne carbon into life forms begins with photosynthesis – the method by which plants take CO_2 from the air, plus energy from the sun and water, and transform them into a carbohydrate used for growth. Different plants, and different parts of each plant, utilise varying amounts of CO_2 from the atmosphere and so have different amounts of carbon in their structure. The woody parts of plants are particularly high in carbon, whereas the leafy parts tend to be made up more of other compounds such as water and nutrients obtained from the soil such as nitrogen. This, incidentally, is why a pine tree makes better firewood than a cabbage – because carbon burns well.

Trees, with their woody tissues, are the classic 'carbon sink'. The carbon already in wood or other plant parts can also be incorporated into the soil, when a plant is dug, composted or mulched into the Earth or when it slowly decomposes on the Earth's surface. The break down of this plant matter is effected by microorganisms such as bacteria, fungi and protozoa, which consume the plant matter and in turn incorporate the carbon it contains into their bodies. As these microorganisms die, the carbon and nutrients that make up their bodies stay in the soil and become available to other microorganisms and plants for growth. Although at every level some of the carbon drawn in from the air is released back through respiration, a sizeable portion of it is held in the plants or soil and so kept out of the atmosphere. [5]

The point is that carbon is always somewhere and is constantly being shifted around. At the moment, too much of it is in the atmosphere, and much more of it could be held in plants and the soil. Many organic practitioners actually define themselves as biological farmers, because they focus on nourishing the microorganisms in a living soil. We may think of soil as an inert object but even a handful of healthy soil can hold billions of microorganisms. In carbon terms, one cubic metre of fertile soil can contain as much carbon as 20 cubic metres of air[6] and it has been suggested that perhaps all the excess atmospheric CO_2 accumulated over the past 75 years could be captured in agriculture soils.[7]

In order to create the healthiest, most carbon-rich soil, the biological organisms in it need constant nourishment. This can occur naturally in forests and on grasslands such as prairies and savannas, where vegetation, insects and animals coexist, each drawing from and contributing to the growing cycle. When people are involved it is easy to disrupt the cycle by harvesting plants and so withdrawing nourishment from the soil, without putting enough back in. Current large-scale agricultural practices use chemical fertilisers to provide the plants with nutrients, but this process does not feed the soil's microorganisms and in fact may kill them.[8] The soil itself is treated as a holding structure for growth, rather than a living entity in itself.

Nourishing the microorganisms in the soil not only means that the soil can hold carbon, but it also creates a fertile growing environment. An optimum level of soil carbon will bring about increased moisture retention, aeration, nitrogen fixation (which reduces the need for nitrogen fertilisers), mineral availability and disease suppression. Carbon compounds also build soil structures that are highly resistant to erosion and the loss of water-soluble nutrients.

In New Zealand, there are many pockets of first class soils (soils most suitable for agriculture) – the land is fairly flat and the soil is deep, friable (easily crumbed) and well drained.[9] These soils represent our easiest soil-based carbon sinks, as well as our most fertile growing land. Intensive agriculture is practised on these first class soils, such as at Pukekohe, on the Canterbury Plains, and in Te Puke, Kerikeri and the Manawatu. However, historically we have tended to disregard soil quality when siting houses and other structures, often wasting good agricultural land. For example, suburbs surrounding the volcanic cones of Auckland, like Mt Albert, sit on magnificent soils.

We have also degraded many of these soils through non-sustainable agricultural practices. At Unitec, students are taken on a trip to Pukekohe Hill. This hill is a major source of vegetables for the Auckland area and beyond. It has huge fields, which contain row upon row of monocrops such as potatoes or onions. Because of practices such as leaving the soil bare between harvesting and planting, there is ample evidence of erosion. The hillside is scarred with pathways where the water has rushed down, taking valuable topsoil with it, which has accumulated at the bottom of the fields and needs to be mechanically re-spread every few years. Fertilisers that feed the crops maintain productivity in the short term while the soil's carbon holding capacity and long-term fertility is degraded.

One of the best ways to preserve our first class soils and to introduce growing capacity to just about any patch of land is to put organic 'waste' into the system. The quantities of organic waste, such as food scraps, that is not returned into our growing systems is incredibly, well, wasteful. Composting involves taking organic material – be it plant or animal waste – and leaving it to be processed by microorganisms (such as those discussed earlier) and macro-organisms such as earthworms, until it can be returned to the soil in a form usable by plants. Composting can be done on a huge scale, if there is the collective coordinated will to return valuable soil food to our farms. It can also be done within communities or home gardens. In New Zealand, most home gardeners are probably familiar with 'cold composting', which is taking scraps from their kitchen, mixing them with weeds and grass clippings, and leaving the lot in a bin until it breaks down. 'Hot composting' is a more efficient, but more demanding, system that involves ideal ratios of the same ingredients as cold composting plus extras such as seaweed, bone meal, paper and fowl manure. The heat comes from the intense microbial activity encouraged by this system. Worm farms are another way of processing organic waste as is Bokashi, a method that ferments food waste in small buckets.[10]

The Beachlands example below shows how composting done by communities can turn what would otherwise be wasted into rich soil.

THE BEACHLANDS MARAETAI RESOURCE DEPOT
by Betsy Kettle

The residents of Beachlands, a small, seaside village about 30 minutes drive from Auckland, were approached by local Permaculture activists in 1998 and agreed to put a composting facility in the middle of their community. The 'Permies' requested the use of the old council works depot for the establishment of New Zealand's first community-based food scrap composting operation. To feed the facility, the Beachlands and Maraetai residents brought their food scraps in 20-litre buckets layered with sawdust.

School children came by the busload and walked on paths through the site. The first path squiggled through the composting operation, past insulated bins in which hot composting of food scraps was going on and then by large worm beds that were fed the raw compost. Leachate (the liquid output from worm beds) and vermicast (worm poo) was harvested. Grass clippings and green waste were also dropped off by residents for use in composting and garden mulch.

The next worm path wound through an organic garden. Children could see vegetable seedlings being grown in potting mix made from vermicast as part of complete recycling system – food used to make soil to grow more food.

Organic gardening workshops were held for adults on how to use grass clippings, vermicast and compost for healthy soil, healthy plants and therefore healthy people.

Besides food scraps, the depot took recyclable glass, plastic, cardboard and metals. As this was the only recycling depot in Manukau, traffic was brisk. Without a shredder, plastic piles built up quickly and a Beachlands trucking contractor helped by hauling bags of plastics into Auckland.

One of the more interesting experiments was borrowing a concrete truck to mix food scraps, paper and shredded green waste. The plan was to use the truck, pre-loaded with shredded green waste, cardboard and paper to run around Beachlands and do a kerbside collection of food scraps. It was hoped that by using a mobile mixing facility compost could be taken directly to farms where it could be composted in modified insulated shipping containers on site. The truck needed to be tested to see if it would discharge the mix quickly enough to be 'economically feasible'. The tests were inconclusive as the concrete bowl was so rough with old concrete that the mix came out too slowly.

This project was before its time – there was insufficient political will to develop the mobile compost truck into a viable operation. Also, kerbside collection of plastic and paper had reduced people's motivation to take food scraps to the composting facility, which led to it eventually shutting down. Today, however, the situation could be very different as communities realise the contribution their food scraps could make to creating a sustainable future.

In 1908, the city of Shanghai sold to a contractor, for $31,000 in gold, over 79,000 tonnes of human bodily waste that was then sold on to farmers.[11] As mentioned earlier, China, Korea and Japan have utilised such waste for thousands of years. Today, we have become so dissociated with our personal waste stream that we defecate into perfectly good drinking water, whilst sitting on a white 'throne' and 'cleaning' ourselves with bleached or embossed paper. While there are numerous challenges in turning what goes into our sewerage pipes into something that can nourish plants, let alone food crops, it is worthwhile remembering that in any truly sustainable system, our bodies must also be involved. If you are brave enough (and your local authority allows it) there are many good composting toilets on the market that make it easy to turn uncontaminated human toilet waste into something that can go in the garden.

FROM INDUSTRIAL FOOD PRODUCTION TO DEEP ORGANICS

As we have already touched on, industrial forms of agriculture are bad for the life of the soil. They also create ecosystems based on a very limited number of species, by practices such as monocropping – growing the same crop over and over again on the same piece of land. These ecosystems depend on fertilisers and pesticides, are disease prone and, as we saw with Pukekohe Hill, are often vulnerable to erosion. A number of analyses have been conducted that compare the effectiveness and sustainability of various food production systems.[12] Table 1 (overleaf) is adapted from Hill's work on redesigning agricultural ecosystems.[13] It divides agricultural systems into four models and draws attention to a number of features – energy use, waste production, use of chemicals, approach to problems, attitude towards productivity versus maintenance, and respect for complex, pre-existing ecostructures – that differ between industrial-based models (shown in columns 1 and 2) and organic-based models (shown in columns 3 and 4).

Industrial production systems see soil as inert, structural and mechanical, with productivity reliant on chemical processes. Weeds are seen as problems to be eliminated by the quickest and easiest means available. Existing ecosystems are ignored. The emphasis is on large 'factory' farms, which are energy hungry, using machinery and factory-produced fertilisers and pesticides, thereby adding considerably to the carbon imbalance in the atmosphere. Industrial models are not based on maintaining the land for generations to come, but on producing as much as possible in the short term. The efficiency model has the same mindset as the industrial, as any change is viewed as an economic imperative. Chemicals are used more specifically, with greater care not to waste them, machinery and fields are larger and productivity is greater. Efficiency farms do have some environmental advantages, such as less fertiliser leaching into waterways.

Shallow organics involves a significant shift in thinking and approach and we are seeing many moves towards this in society today as demand for certified organic produce grows. 'Safe' chemicals emerge and are used in production, weeds are seen as important to the health of the environment, and the biology (and chemistry) of soil is recognised, with the emphasis being on feeding the soil and not the plants. 'Weeds' are often encouraged. Carbon in soils increases.

However, it is not until we take a deep organics approach that full recognition is given to the complexity of ecosystems and of the

INDUSTRIAL	EFFICIENCY	SHALLOW ORGANICS	DEEP ORGANICS
Least sustainable	First steps towards sustainability	Second steps towards sustainability	Most sustainable
Narrow focus – farm as factory	Efficient factory model	Softer factory but still input-output focused	Broad focus, farm as complex ecosystem
High energy, high waste	Conservation of energy, some reduction in waste	Conservation of energy, some low-energy technologies introduced, less waste	Low power and no waste
Chemically based products (fertiliser, pesticides) and biotechnology	More finely tuned chemical products (e.g. slow release fertiliser)	Recognition of soil biology and use of 'natural' chemical inputs	Self-sustaining with no need for chemical inputs
Problems, such as weeds, are seen as enemies to eliminate and control directly with products	Efficient control, but problems still seen as the enemy	Natural controls, with greater sense of problems as 'I am not managing this well'	Problems seen as imbalances, indicating are changes needed to the system
Maximise production (neglects maintenance, and no focus on future generations)	Maintain production while improving maintenance. Some focus on future generations	Improved maintenance and attention to sustainability and biodiversity	Optimise production. Emphasis on maintenance and meets real needs of present and future generations
Infrastructures dominate that damage pre-existing ecostrucures (riparian zones, wetlands, soils, etc.)	Infrastructures have lower impact on pre-existing ecostructures, but still unacceptable	Infrastructures exert little or no damage on pre-existing ecostructures	Infrastructures designed to enhance eco-structures

Table 1 Comparison of agricultural strategies

desirability of designing a growing world that is self-sustaining. Andre Domine captures the essence of deep organics (which we will simply call 'organics' for the remainder of this chapter) by explaining that:

To use the word 'organic' betokens an attitude of respect which acknowledges that plants, animals and human beings are all part of the same system of life. This interconnectedness of all living beings determines their interdependence upon each other for better or worse. 'Organic' is therefore not a label which can be superficially used to decorate carrier bags, boxes, tins, or bottles, but is rather an attitude to life. It is only when farmers, gardeners and winemakers as well as officials, inspectors and traders allow themselves to be motivated in an appropriate way that the term organic will become and remain a seal of quality and a guarantee. Only if consumers give active and consistent expression of their desire for organically grown foodstuffs will the ground be prepared in a literal sense – upon which such products can flourish. Once the understanding is there that the Earth is alive, that it represents a multifaceted, complex co-existence of an innumerable number of living beings, without which there can be no healthy growth, it is only then that organic cultivation can begin.[14]

Organic farming has also been described as like surfing[15] or jazz[16] – a creative, responsive process, rather than a rigidly defined set of practices, formulae, procedures or activities. The emphasis is on optimisation – production that retains and enhances soil health and can be maintained indefinitely. Water leaves the land cleaner than it enters. Food is packed full of nutrients that keep us healthy, and food waste goes back to where it came from to further enrich the soil. Children are brought into the produce growing process, learning how to design and build living systems as well as how to manage and maintain them. In this way, not only productive land, but also knowledge is passed on to the next generation.

The beauty of organic systems is that they can be used anywhere by anyone and they can be adapted to suit our current agricultural challenge, which is to maximise carbon capture and storage in balance with greatest ecosystem health. Much of this can, and is, being done in rural areas. Farmers are planting carbon-rich trees on their land and retiring land not suited to agriculture, and there are numerous examples of thriving organic farms and production sectors (kiwifruit, pip fruit and dairy) shifting significant portions of their enterprises to certified organic.

All these efforts are hugely important, and buying organic produce is one way of supporting sustainable rural agriculture. However, we can also use organic principles to turn our cities and towns into living ecosystems that both pull carbon from the atmosphere and allow us to grow food near where we live – a double whammy for carbon neutrality as any food plucked from your backyard or grass verge avoids the food miles involved in transporting it from the farm to your table. The use of organic principles is a key part of a carbon-conscious shift in society.

TRANSFORMING URBAN AREAS
When the Soviet Union collapsed in 1990, Cuba's access to oil-based fertilisers and pesticides was dramatically reduced.[17] At the time, it was a highly industrialised farming nation, using more fertiliser per hectare than the USA. With the loss of these products, a radical rethink was needed in order to feed the nation's population of 11,000,000. Survival agriculture took over, and almost every piece of arable land was planted, not just in the countryside but in the cities too. Community cooperation was essential as wasteland and even rubbish dumps were converted into gardens. Within about five years, 50 per cent of the food required to feed the capital city of Havana was grown within the city itself. In 2006, Cuba used 21 times less pesticide than before the crisis and 80 per cent of its food production was organic.

The Cuban response to their oil crisis illustrates how quickly it is possible to re-adapt to organically based agriculture, largely grown in and around urban human settings. The question is whether we can adapt in the same way to our climate crisis (and, as a by-product, be better prepared to withstand the consequences of peak oil).

To begin we need to reclaim land for growing within and on the outskirts of our urban areas. Special attention should be given to retaining land with first class soils for agriculture – rather than covering it with shopping malls, transport corridors and more houses. New subdivisions could be required by local authorities to have a set proportion of planted land relative to the area covered by houses and other structures such as decks. Planted areas are likely to be far more important in the future than the currently compulsory car parking space. Properties could be arranged to create large growing areas, not small, isolated backyards. This could be achieved by having groups of eight or more houses arranged around a common growing area. The resultant vegetative web allows for cross pollination of fruit trees and the interaction of diverse plant and insect species needed to create strong, self-sufficient ecosystems. In established suburbs, it is still possible to partially create this effect by simply planting wherever possible, and creating living edges between properties. If you need a fence consider whether a rosemary hedge, or a raised boxed garden bed planted with trained raspberries or blackberries might work. Who knows, if Kyoto-style carbon costs are passed on to local authorities, in the future the rates paid by a property owner may be inversely proportion to the amount and type of planting on the site. Satellite photography and geographical information systems would make this relatively easy to measure.

As already noted, a healthy ecosystem has a huge range of species providing each other with food and shelter. Many of the plants commonly thought of as weeds, such as Queen Anne's lace, ragwort and fennel, are habitat to beneficial insects that serve to pollinate fruit trees or control the insects that compete with us for the vegetables we plant. Weeds can simply be left to grow along railway tracks, highways and median strips, and managed appropriately rather than be sprayed by the local authority in order to achieve a tidy look.

Diversity of food crops is an important principle in any human community, and increases food security. If New Zealand is limited to only eight varieties of apples or potatoes, this greatly limits the gene pool and increases the risk of pathogens that can decimate our food supply. Seed saving and planting heirloom varieties are important ways of insuring this diversity. For example, Koanga Gardens is an excellent source of heirloom seeds.[18] Urban gardens are in many ways ideal sites for ensuring the survival of diversity, as a home gardener can grow several varieties of carrots (it is pretty cool to dig out a handful of yellow, purple and white carrots along with same old orange ones) and every house on the street could have a different species of apple tree suited to the region, soil type and conditions. As a consequence, the pressure to plant, harvest, distribute and sell food through the mechanical and low-labour means that encourage uniformity of product on large industrial-style farms would disappear.

In anticipation of, or as a consequence of, reduced vehicle use, many long concrete driveways could be removed and replaced with grapevines or beans. Even now, rows of potatoes or feijoas could be growing on verges between the footpath and the road. Large fruiting trees, such as avocados and walnuts, could be planted in parks and golf courses, and smaller fruit trees such as lemons, apples and cherries could line city streets and school playgrounds. Trees with a particularly high capacity for capturing carbon should also feature. Hardwoods native to the local area are especially good for this purpose. Not only are hardwoods optimal carbon sinks (they grow especially carbon-dense wood), but local native varieties are most likely to survive, being already genetically adapted to your region's environment.[19]

When planning how to maximise your own property, it is important to think about what land you have available, its aspect and other features such as drainage. For example, do not plant large trees on a 400-square-metre section that will block out sun and make it difficult to grow vegetables (or have solar panels). Keep your sunny areas for growing produce and put the compost, worm farm, water tank and bike shed (or car) in the shade. If you need shelterbelts, you can use nitrogen-fixing trees like tagasaste for soil improvement or insect-attracting trees like bottlebrush. Include some quick-growing carbon crops, such as broadbeans or corn – after harvesting, the residue can be dug into the soil, where it feeds those carbon-hungry organisms, contributing to both carbon capture and soil health at the same time.

As our towns and cities transform into complex, diverse and hardy ecosystems pulling in CO_2 from the atmosphere and providing us with fresh food that can be obtained without a trip to the supermarket, we will need to feed food and other waste back into the system. While much composting can be done on individual properties, local authorities will need to develop an infrastructure to process the rest. Community-based and mobile composting facilities, such as the Beachlands initiative discussed earlier, are one solution. These will be particularly important for apartment buildings and commercial enterprises, including – but not limited to – restaurants. One day, we may even be able to safely process and re-introduce the human waste that goes into centralised sewerage back into our agricultural systems.

National and local government will need to provide education and support to home gardeners through community centres, schools and demonstration homes (see text box below). Schools could also function as demonstration communities through the Enviroschools programme discussed elsewhere in this book. This would serve the additional purpose of teaching children about soil and growing.

In the text boxes below are two examples of urban properties that have used many of the principles discussed – and others. Betty Kettle's suburban Auckland demonstration home shows how it is possible to grow an incredible variety of food on just a 650-square-metre section. An organics pioneer and an embodiment of its lifestyle is former Lincoln University senior lecturer Bob Crowder. He describes developing his Christchurch property along the same lines as the organically-based Biological Husbandry Unit he was part of at the university.

PAKURANGA'S URBAN PERMACULTURE DEMONSTRATION HOME
by Betty Kettle

Auckland's 'Pakuranga urban Permaculture demonstration home' had an inspirational effect on Auckland's suburbanites. Between 1996 and 2001, over 2000 people visited this fifth-of-an-acre section. The open day tours traced a looped path around the property, passing information boards along the route that explained the different features of a sustainable garden. The property was a very average 1960s one-storey wood house with a corrugated iron roof and garage. It is likely that its ordinariness inspired others to think about what they could do themselves at home for very little money.

The tour started in the front yard, where the garden featured an earth wall topped with a split bamboo roof. Workshops on earth building and bamboo construction were held here by local experts, with the community invited to attend. 'Using local, abundant and biodegradable materials' was the theme of these sessions. Many other workshops had participants design, build or grow various features of the garden. Regular workshops were sponsored by Manukau City Council on worm composting, hot composting and organic gardening in an effort to reduce organic waste to landfill.

Of particular interest was how to prepare garden beds using massive amounts of grass clippings to shade out weeds and improve the soil structure at the same time. Breaking down clay soils by adding lime and gypsum was also eye-opening knowledge that was much appreciated, as many people thought such soils were 'too much work to be bothered with'.

The route carried on around the side of the house to a narrow, vertical garden planted with subtropical trees. The information board explained how being able to identify and use microclimates enabled home owners to make the most use of all available space, to grow intensively with a great diversity of edible plants. Growing plants in containers, the herb spiral and the cutting garden were also given as examples of intensive, space-saving ways to grow food in small areas.

Over 50 species of fruit and berries were planted in the 650-square-metre section, providing a year-round supply of small amounts of fruit. For instance, cherimoyas and casimaroas in the subtropical orchard produced fruit in spring when other fruit trees were just starting to flower. Ways to pack in the fruit trees through pruning, growing on dwarf root stock, using as ornamentals, planting on the verge (and in the chook pen, in pots and as hedges) were demonstrated.

It was probably the watering system that provoked the most political change. At that time, Metrowater was discussing expanding the Auckland Region water supply from the existing dams to include pumping from the Waikato River to the south. The rainwater-collecting system demonstrated how using rain tanks for non-potable use (toilet flushing, laundry and gardening) could cut water consumption by 50 per cent and not touch the existing internal plumbing. It was also explained that having rainwater tanks provided emergency water in case of earthquakes that could potentially disrupt city mains supply. Visits by Waitakere and North Shore City Councils eventually prompted changes in by-laws that promoted rather than prohibited urban rainwater tanks.

Other features of the garden were: three different composting systems using worms, hot composting and passive piles; improving soil depth using no-dig sheet-mulching gardens (using more grass clippings); keeping urban chickens; keeping urban beehives; using guinea pigs as lawn mowers; on-site storm water management and intensive four-season vegetable harvesting. The idea of eating the guinea pigs was not seen as being fashionable.

Here is just one testimony to its success. It was extremely easy to get grass clippings from landscape contractors in the region before the project began. Five years later, however, the local contractors apologetically said that all their grass clippings were now being used by others!

MY GARDEN
by Bob Crowder

When my rental accommodation burnt down in the 1970s, there was nothing to replace it until a special friend showed me a little colonial house on a big wild section where poplar trees towered into the sky and azalea and lilac blooms cascaded fragrance into the drive. I was sold and entered a personal paradise.

Looking back, it is difficult to comprehend how I managed everything in that era as the organic movement entered its most exciting and controversial era, with the Biological Husbandry Unit [BHU] at Lincoln University constantly expanding. However, we were all learning and as each year passed we learnt that with the correct respect, nature had a way of looking after itself, and that making the system tidy for our eyes was not necessarily the best strategy.

My own 2500-square-metre paradise in urban Christchurch followed the very same principles being developed at the BHU with a view to creating a personal retreat that was as sustainable as possible. As mentioned, it had massive poplars when I bought it but also, at the front, some delightful plantings of trees and shrubs already under-laid by masses of blue bells. The poplars were heavily coppiced but not bulldozed and to this day are managed each year for firewood which is augmented with wood from plantings of various willows and some eucalypts. Nothing has left this property in over 26 years.

Fruit self-sufficiency was another goal – why have an ornamental cherry when a Dawson and/or a Stella is just as gorgeous in spring? What is better than the spring display of plums, apples, pears and nashi? And the persimmon in autumn must have the most flamboyant colours of any tree. Then there is the majesty of the sweet chestnut and the walnut and the kiwifruit – the last being the most rewarding of any plant in the garden for providing that valuable winter vitamin C fix.

At last count, this 2500-square-metre plot of urban living had some 30 different types of fruit and nut available, all producing quality produce and, through careful selection of varieties and organic growing, receiving nothing other than a bit of seasonal pruning.

Beneath this fruiting canopy, there is room for a rich understorey of spring flowers and winter self-sown salads of all descriptions. Out beyond the drip line of the trees, there is room for the more sun-loving vegetables and herbaceous flowers and herbs, all in glorious chaotic splendour with only just a nudge of management in the right direction.

All in all, a very satisfactory situation that leaves plenty of time to sit on the bench under the spreading chestnut tree with my afternoon tea and contemplate this little bit of environmental paradise.

AOTEAROA/NEW ZEALAND AS AN ORGANIC ECO-NATION BY 2020

By 2000, as President of the Soil and Health Association,[20] and lecturer at Unitec New Zealand I (Brendan) was part of a small group who initiated the vision of New Zealand being an organic eco-nation by 2020. We published visions, timelines, stimulated debate and held a series of national conferences and think tanks, the most recent of which was in May 2007.[21] The concept is one that places New Zealand as a world leader in organic consciousness, practices and technologies. We continue to believe it is not only plausible and possible, but one of the more realistic visions for the future.

We are not alone on this. Numerous international organisations and institutes are calling for a similar vision. For example, an international organic umbrella group, the International Federation of Organic Agriculture Movements [22] with over 800 members from 110 countries has a mission no less than the 'Worldwide adoption of ecologically, socially and economically sound systems that are based on the principles of organic agriculture.' Publications, interviews, and supporting concepts have flowed from our 2020 vision. Besides offering organic certification, our work has generated music CDs,[23] politicians and business people have been influenced, policies created and commitments made. Tangible successes have been apparent, including a coordinating body, *Organics Aotearoa/New Zealand*;[24] a new scientific journal, *Journal of Organic Systems*,[25] local 'Regional Food Economy' programmes, and the development of GE-free regions throughout New Zealand. Initial analyses suggest that this vision could also hold considerable economic opportunity.[26] People are keen and the government is beginning to listen.

In summary, we believe that the concept of an eco-nation by 2020 is:

- An achievable vision for a sustainable future
- A real opportunity to secure economic security in the 21st century
- A method by which our nation can focus on a common purpose
- A means of creating meaningful partnerships between citizens, communities, consumers and government
- Based on sound environmental land practices
- A place where we want to raise our children.

It requires:

- People and communities to participate in the redesign of the ecosystems in which they live
- Incentives from local and national government
- Continued research into appropriate technologies and methods of action
- Immediate appropriate action
- Education at all levels to encourage people, communities, consumers, business and government to adapt to appropriate principles and practices.

The award-winning Unitec Hortecology Sanctuary – Mahi Whenua – arose from the organic eco-nation vision. It is a showcase for anyone who wants to learn about organic principles and how to grow food in an urban setting.

> **DESIGNING OUR FUTURE**
> **The Unitec Hortecology Sanctuary/Mahi Whenua**
> The certified and award-winning Unitec Hortecology Sanctuary / Mahi Whenua (UHS) was established in 1999. It is home to the Pacific Centre of Sustainable Communities. It was inspired by a will to demonstrate a sustainable future that combines biodiversity, conservation, food production and waste management into a single experience. Inspiration came from the founders' (Brendan Hoare and Richard Main) wanderings through Asia and personal commitments to an organic future.
>
> Located on the banks of Oakley Creek in Unitec's Mt Albert campus, the Sanctuary builds on the foundation of learning by doing. It owes its existence to the hundreds of students who have sweated, laughed, bled and toiled in its creation.
>
> The results have been inspiring. It has been a centre of learning and extension for the Auckland region, nationally and at times internationally for the last eight years. Both staff members now play important roles on the local, regional, national and international sustainability stage. Having the UHS as a physical manifestation of the future they espouse is a powerful tool.
>
> The 1.5-hectare UHS site was transformed from a plot of urban grass to a demonstration of a whole-systems approach, with appropriate land and water use by utilising swales, biodiverse

shelters, home gardens, extensive food production and rotations. By far the most popular component is the food forest, where over 50 rare edibles have a home.

The food forest is now reaching maturity, with the larger tree crops like avocado breaking through the canopy of nitrogen-fixing pioneers while second tier species like babaco, mountain pawpaw, feijoa and pepino reach maturity and drip with fruit. An understorey of herbs, flowers and fungi blanket our forest floor while the forest canopy is fringed with sun-loving citrus.

The only liquid that touches the leaves of the plants is water. No chemicals, not even organic certified ones, are used. Waste recycling using a range of compost methods, seed saving and co-operative action and sharing of resources helps demonstrate how infrastructures (human-made systems like swales) can enhance ecostructures (naturally occurring ones like water courses) through good design.

The UHS is not the only transformation. Students, staff and members of the public have been able to experience what a sustainable food-producing environment actually means. As a result many have been able to transform their lives, home and work environments and livelihoods. UHS has been a place that has enabled leadership to emerge, and for students to feel confident enough in the art of organic systems approaches to land management to redesign their own futures.

It is a focal point for raising awareness and a centre of inspiration, motivation and the development of a caring attitude for urban and peri-urban environments. It is also now a place of celebration, hosting numerous launches of visions (2000 Organic 2020 Conference), technologies (Vertical Compost Unit), community projects (Maori potato exchange) and national projects (Organic Farm New Zealand). These have been carried out with extraordinary commitment from the team and support from an enabling campus.

Future plans include a one-stop enhanceability shop that integrates a range of skills and expertise into a single experience of what our immediate future may look like.

CONCLUSION

Tim Flannery describes the human species as 'future eaters', production and consumption gluttons with little or no consideration for future generations.[27] His description challenges us to consider if we want to be known as the generation who knew what was coming and ignored the signs. What sort of world do we want to leave our children? By growing carbon-rich soil, trees and food in not only our rural, but also our urban areas, we can start to reverse the accumulation of greenhouse gases in the atmosphere and create a viable future for the people who will come after us.

We can all take part in this process, by planting trees, growing food organically (even if it is just sprouts on the window sill), making sure our food scraps are composted, joining one of the existing organic organisations, and supporting other people's efforts by buying organic when we can. In today's society, growing organically is one of the most radical acts any individual and or family can do. You will not only be caring for yourself, your family and the neighbours, but also for the global environment, as you will be helping grow the healthy soils that attract and hold carbon that would otherwise be warming the planet. Creating a viable future starts with all of us.

ENDNOTES
1. http://www.fiu.edu/~fcf/marti.pensamientos.10797.html (Accessed 3 Oct 2007).
2. King, F.H. 1927. *Farmers of Forty Centuries*. London: Jonathan Cape. Personal communication with Professor Chen Shih-Shiung, Dean of General Affairs at National Chung Hsing University, Taiwan.
3. For more information, see *Achieve Consistent, High Standards of Environmental Performance for Waste Treatment and Disposal* at http://www.mfe.govt.nz/publications/waste/waste-management-nz-oct05/html/page3d.html (accessed 28 Sept 2007).
4. Yeomans, A.J. 2005. *Priority One: Together We Can Beat Global Warming*. (p. 88). Arundel: Keyline Publishing. Ordering information is available at http://www.yeomansplow.com.au (accessed 28 Sept 2007).
5. McLaren, R.G. and Cameron, K.C. 1996. *Soil Science*. 2nd Edition. Melbourne: Oxford University Press. This contains a description of the carbon cycle.
6. Yeomans, A.J. 2005.
7. Yeomans, A.J. 2005. See chapter 5 and pages 87—106.
8. Lampkin, N. 1990. *Organic Farming*. Ipswich: Farming Press Books.
9. McLaren, R.G. and Cameron, K.C. 1996.
10. To learn more about Bokashi, visit www.bokashi.co.nz (accessed 28 Sept 2007).
11. King, F.H. 1927.
12. Odum, E.P. 1969. The strategy of ecosystem development. *Science 164*, 262–270; Potts, G.R. and Vickermann G.P. 1974. Studies on the cereal ecosystem. *Advances Ecological Research 8*, 107–197; Altieri, M.A. 1995. *Agroecology: the Science of Sustainable Agriculture*. Boulder: Westview Press; IT Publications.
13. Hill, S.B. 1992. Environmental sustainability and redesign of agroecosystems. EAP Publication 34. Montreal: Ecological Agricultural Projects, McGill University. http://www.eap.mcgill.ca/publications/eap34.htm (accessed 28 Sept 2007).
14. Domine, A. 1997. Forward. In *Organic Wholefoods: Naturally Delicious Cuisine*. London: Konemann.
15. Haikai Tane (personal communication, 2000).
16. Jamie Taite-Jamieson (personal communication, 2001).
17. This description is taken from the film *The Power of Community*, directed by Faith Morgan and released in 2006. Cuba: AlchemyHouse Productions.

18 http://www.koanga.co.nz (accessed 28 Sept 2007).
19 Olmec Sinclair, carboNZero (personal communication, 2007).
20 Founded in 1941. For more information, visit http:/www.organicnz.org (accessed 28 Sept 2007).
21 See the Soil and Health Association's magazine *Organic NZ 683*.
22 For more information, go to http://www.ifoam.org (accessed 28 Sept 2007).
23 For information about the Econation 2020 campaign, go to http://www.econation.org.nz (accessed 28 Sept 2007).
24 http://www.oanz.org.nz (accessed 28 Sept 2007).
25 http://www.organic-systems.org.nz (accessed 28 Sept 2007).
26 Saunders, C.M. 2000. *Markets Signals and Problems*. Invited Paper to Plant Protection Society Conference on Plant Protection Problems and Organic Products. Christchurch, August 2000.
27 Flannery, T.F. 1997. *The Future Eaters: An Ecological History of the Australasian Lands and People*. Sydney: Reed New Holland.

Back to the drawing board:
sustainable design

David Trubridge

David Trubridge is one of Aotearoa/New Zealand's leading furniture designers, with an international reputation spanning all continents. He designs and produces his pieces with his large team at Cicada Works in Hawke's Bay, from where they are exported all over the world. He exhibits every year at major shows in Milan, Paris, London and New York. His work, which displays a deep environmental consciousness and responsibility, has been described as trend setting by Terence Conran and design editors. David teaches and lectures about sustainable design around the world.

THE CURRENT SITUATION
Our environment has been degraded by pollution and the profligate use of fossil fuels, both of which are inextricably linked to rampant consumerism in the Western world. People are also suffering – whole populations are denied resources and relegated to poverty. As a particularly cruel twist, it is the poor countries that are likely to suffer most from the rising sea levels and increasingly severe weather – all effects of the global warming caused by the rich. Designers are complicit in this process by seducing the consumer into buying new things every year, not because the consumer necessarily needs them but because the whole commercial set up, from manufacturer to shipper to retailer, needs to sell them to survive. In the process, irreplaceable minerals and resources are strip mined for a brief year or two's use before being dumped on a landfill, and countless tonnes of carbon are unnecessarily released into the atmosphere.

THE VISION FOR 2020
We will take only our share of resources and live in a way that can continue forever. We will fulfil our needs in such a way that we do not jeopardise the needs of others on the planet, or those of future generations.

STRATEGIES TO ACHIEVE THE VISION
Everything has to be redesigned. Designers must find new and better ways of doing things, so that we can all have an acceptable lifestyle. The role of design will need to revert, from helping to cause the problem, back to ensuring our survival, as it always used to. For example, objects will need to be designed so that every component can be separated and used again, indefinitely. But more importantly, all our governing and finance systems have to be redesigned, because presently they are failing us. Current forms of communication must be redesigned to create global networks that will circumvent politics and big business. As a consequence, we will be better linked globally but more concentrated locally. World trade of ideas and intellectual property will replace physical commerce, so that objects can be sustainably made with local materials and locally produced energy.

INTRODUCTION
2006 saw a remarkable change in the world's attitude to climate change. 'Tipping point' is a phrase that is often used in discussions on climate change. It is the moment when an irrevocable series of chain reactions is initiated, in this case weather changes being brought about by our

actions. Public opinion went through its own tipping point in 2006, thanks largely to Al Gore's intervention[1] and subsequently the Stern Report into climate change.[2] Suddenly, climate change became a fact and the arguments were largely over. The media were swamped with news and discussion on how to solve the crisis and we were offered a valuable opportunity to reappraise our lifestyles. How do we make the most of this opportunity in a positive way? This chapter is not a detailed scientific report; it is an overall philosophical look at the role of design – past, present and future – in our survival. It comes from my experience of designing things and from observing how design functions in human societies.

If, in 2006, human-induced climate change became a fact for most New Zealanders, people living on the edge of survival have known about it for many years. Their knowledge is not based on scientific records, but on an intuition built from generations of living in harsh marginal conditions where they had to acquire an empathy with the patterns of seasonal change. The Inuit and Eskimo of North America have watched the sea ice recede, the glaciers melt, the migration patterns of creatures change and the vegetation of the tundra be transformed. For as long as they have lived, their survival has depended on a finely tuned balance. In temperate climates like we have in Aotearoa/New Zealand, survival is relatively easy, and temperature changes make much less difference. But for places like the Arctic or semi-desert regions, where survival is a harsh reality, one or two degrees of change is crucial.

Science, however, cannot be based on intuition. For years, scientists were unable to turn their data into a convincing model of future weather patterns that proved the effects of global warming. The weather is such a tumultuous mass of turbulence that not even the most powerful supercomputers can build models of absolute prediction. The number of possible outcomes from a single weather event increases exponentially so rapidly that predicting even one week ahead becomes unreliable. Because definitive proof of future harm from current activities has been impossible to provide, for many years, politicians refused to act. They were influenced instead by those who had most to lose, such as oil companies and automobile industries protecting their short-term profits. Governments and businesses exploited the scientists' lack of absolute proof for as long as they possibly could, just when we needed to be acting and planning for the future.[3]

DESIGN HISTORY

How have the Arctic peoples survived for so long in such an inhospitable environment? My answer is simply because of design. In fact, the same applies for the whole of humankind's survival on this planet. As a species, we are physically ill equipped to survive. We lack powerful claws and teeth or the athletic abilities of speed and strength. But, we have a very smart brain. We are able to use it to do things that other species cannot. We have used our brain to design tools for survival.

Before I go further, I should clarify my use of the word 'design'. We are usually presented with the words 'art', 'craft' and 'design' as descriptors or categories of objects or creator. This leads to all sorts of confusion and argument because I believe these are really words that describe processes, not things. And everything that is made uses a mixture of all three processes. In creating any new thing, the maker works through the art process, the design process and the craft process, although the balance between the three will vary considerably. A Renaissance painter evenly balanced all of them: first, as an artist in the greater vision of the work – the ideas behind it and how it changed perceptions; then, as a designer in giving the ideas form through composition and layout; and, finally, as a craftsperson in the physicality of mixing pigments, preparing the grounds and applying the paint itself, all in a way that ensured the artwork endured. And there is a design component in the process of creating an object just as much as there is in creating an abstract system, such as a ritual or even a system of accounting.

For most of my life, I have been a designer of some kind. I was trained as a naval architect, or boat designer, in northern England, but a lifestyle choice found me designing and making furniture soon after graduating. For a while, I also worked as an architect. I have always loved sculpture and my furniture often has a strong art content, where I give as much weight to the aesthetic function of an object as to its practical use. But I do try to keep that Renaissance balance of three equal parts, rather than allowing art, craft or design to dominate the process too much.

So, when I use the word 'design' I do so to indicate the process of thinking out a solution to a problem. In the Renaissance example, the painter had an idea to express – how could that be best done? Or a simpler example: when the earliest primitive people lived in caves, perhaps there came a moment when someone thought to put a slab of stone onto supports to lift it off the ground, thus making a table. Maybe this was one of the first acts of design. Food on the ground was probably being wasted because it was trampled by children and animals. Food was an expensive commodity that required many hours of dangerous hunting or laborious collecting, plus preparation and cooking. Maybe the first designer saw this problem and thought about how he or she could make things better. By reducing food wastage, the designer had improved the tribe's chances of survival. Thus, the role of the designer was established.

As human cultures slowly developed, designers played a crucial role in ensuring our survival by constantly thinking about, and improving, conditions. They gave us clothes to keep us warm in the cold, and houses to shelter us from the weather. They created boats for travel and for fishing, as well as such items as food traps or storage baskets. All this was necessary because humans were weak and physically ill equipped to survive, while the environment we lived in was hostile and powerful. Unlike other creatures, we survived because of our design ability more than our physical prowess or attributes.

Sometimes designers were cleverer in their thinking. In the Corsican Stone Age village of Filatosa, I saw large stone menhirs with carved faces, which may be the earliest example of such figurative carvings. They are thought to have been placed around the village to scare off possible attackers, perhaps by persuading them that these superhuman giants protected the inhabitants. In a different more subtle way, the designer was still helping to ensure the survival of the tribe.

Another example of such thinking is a carved figurehead on the prow of a canoe from Vanuatu. It is a stylised frigate bird that elegantly completes the long forward sweep of the canoe's lines. The beautiful bird sits poised, expectantly gazing ahead, every simple detail perfectly expressing its purpose. However, it is not just an attractive form – it has a purpose. Frigate birds spend their day flying over the ocean looking for food, but they always find their way home to land to roost for the night. So too the islanders; this bird symbolises a safe return for the fishermen out in their canoes. With its unerring instinct, it will guide the canoe home with its precious catch of food. Here the designer not only created the boat for fishing, but also imbued it with the magic required to ensure both the survival of its occupants, and the tribe that are dependent on the food they provide.

Over the millennia, human endeavour has come to encompass more than just survival. Originally, humans were supplicants to nature, mitigating its power over us through religion. We have developed deeply felt beliefs, religions and superstitions. These beliefs have given us more confidence of our place in the world and hence a greater survival instinct. Throughout history designers have found ways to reinforce those beliefs by creating holy statues and paintings. In doing so, they have bequeathed to us some of the pinnacles of human genius in the great cathedrals, temples and religious paintings.

During the early period of human development nature was strong and we were weak, but we were able to live in a balance with our surroundings. This is best illustrated today by the house of the Chencha people in Africa. It is in effect an upside-down basket woven from bamboo. The vertical sticks are placed in the ground and horizontals are woven through them. The whole is covered with 'tiles' made with the papery part from the top of the bamboo. In local currency, these structures are expensive and it takes a long time to save up for and to collect enough for a house. In fact, it takes about 20 years. By this time, the bamboo at the bottom of the existing house has rotted and the house has slowly sunk into the ground. However, in that time, enough materials have been collected to build a replacement. This is sustainable design: materials are used only as fast as they are replenished and life can continue in this balance indefinitely.

DESIGN TODAY

All that changed with the Industrial Revolution. Suddenly, the means of multiple production became available and that radically altered the role of designers. For example, highly intricate and delicate forms that would have taken countless hours to cut and shape individually by hand, could now be made in large numbers out of, for example, cast iron. Instead of creating things to ensure the survival of their immediate social group, designers ensured their personal survival by producing whatever goods they could for others to buy. So, the emphasis changed from survival to selling. Community objectives were replaced by individual objectives, altering the whole structure of society. As prosperity grew so did the number and variety of goods that could be sold. The principal justification for making something now was that it could be sold.

In our day, designers rarely work alone. Most work for large companies. These companies constantly need new products to sell – not primarily because the buyer needs them but because the company needs to sell them. The company's commercial viability depends on it bringing out something new each year so that the buyers keep on buying. Therefore, the role of the designer is now to create a new look, a new styling or a new fashion. If the product is too durable in its manufacture, or too timeless in its look, then producing it is not in the best, short-term commercial interests of the company. On top of that, there is an enormous advertising industry employing graphic designers, image-makers and wordsmiths whose roles are to persuade us to buy these things (which we probably do not need). Cell phones are a typical example, being replaced on average every 18 months despite the fact that they can be made to last ten times as long. Getting the consumer to replace his or her cell phone is done by constantly adding new gadgets and functions, and changing the look – not all at once, but incrementally. In the process, energy is consumed, carbon dioxide is pumped into the atmosphere and valuable minerals are irreplaceably lost to landfills.

Advertising pressure makes you the buyer feel inferior and breeds dissatisfaction if you do not have the latest version of whatever. An abundance of glossy magazines support this superficial culture, each one consuming, on average, a quarter of a litre of oil in its production. This is a radical reversal of the role of the designer and raises the question: who is the designer now working for? It is clearly not the consumer any more, whose needs can usually be fulfilled far more simply. Products can be made to last for far longer than is generally allowed or is possible under our current commercial model. There would be no problem with this if it were done in a sustainable way like the Chencha house. But it is not, and irreplaceable material and energy resources are being strip mined and discarded, solely in the name of short-term business profit, with none of the long-term costs (such as those associated with full clean-up or replacement) being paid. Both designers and consumers are complicit in this extravagant waste. Far from design being a tool for survival, it has now become a threat to survival on a massive scale.

'Of all the trillions of tonnes of materials we make, move, transform and manufacture with, as little as 1 per cent of them are still in use in products six months after their sale.'[4] The rest ends up as landfill. When you consider how much material does last longer than six months in the form of buildings and cars, then you realise just how much stuff has a very much shorter life than this. Most of what ends up in the landfill is the unnecessary packaging and throw-away implements whose life is often measured in minutes after purchase.

These results come from a rampant consumer binge that is rapidly using up non-renewable resources and in the process causing serious pollution. The true price of extracting these resources is not being paid by current consumers, but being irresponsibly passed on to unfortunate future generations who will have to live with the mess and the exhaustion of oil and minerals.

Since the Industrial Revolution, roles have been reversed: nature is now weak and we are strong. Collectively we have gained the power to change the face of the Earth and its weather systems. Instead of facing up to this responsibility, we are riding roughshod over the Earth and many of its peoples. We are squandering the future with our short-term greed and out-dated Industrial Revolution attitudes.

The consequences? 20 per cent of the world's population consumes 75 per cent of the resources.[5] It is this privileged minority that is doing all the damage – causing pollution and global warming. What about the other 80 per cent of the people? They also have a right to our lifestyle and luxuries, and some countries like India and China are now claiming it. But to bring that 80 per cent up to our present standard requires a four-fold increase in production[5] at the current population level. However, by the time these countries reach Aotearoa/New Zealand's standard of living, in say 50 years, their populations may have doubled, which will instead require an eight-fold increase in production. This scenario is clearly impossible – so much damage is being caused by so few of us already, and the resources have finite limits.

But it gets worse than that. Our economic model requires growth, ideally at least 3 per cent per year. If the economy is not growing, people start to get worried and talk of a recession or even a depression. We are told we must buy more. So, if we factor in a modest 3 per cent growth, the six-fold increase becomes a staggering 34-fold increase in 50 years time (4 per cent growth requires a 55-fold increase).[5] So, we are faced with two choices:

1. If all present and future populations are to share the world's dwindling resources evenly and fairly then we must accept a drastic reduction in our consumption, or,

2. If we maintain our current levels of consumption, then we must keep the remaining 80 per cent of the world's people in poverty, which is morally untenable. And we will exhaust all main resources within 50 years.

How do we change? How do we find a way to satisfy the needs of today, without jeopardising the needs of tomorrow?

DESIGN TOMORROW
There are two ways of looking at a serious situation: it can be considered a threat or an opportunity. Seeing it as a threat is a negative, knee-jerk reaction more doomed to failure because you are on the defensive. Seeing it as an opportunity is, by contrast, a positive reaction, which takes control and looks to the future.

I have shown how historically the role of designers has been to aid survival. More recently, that role has been reversed to become a threat to survival through the production of waste and pollution, and by causing global warming. Now, it can and must be turned back again.

Just as design allowed us to survive in the past, so it can do so again now. This is a time of unparalleled opportunity and need for designers. For the human race to overcome the threat of global warming, surely we have to redesign everything?! The process does not start with objects but with the way we do things. This ranges from worldwide political and financial institutions right down to the most mundane personal ritual. Our current systems are failing us. Manufacturing is based on 200-year-old models from the Industrial Revolution that are reliant on abundant cheap energy and materials. Politics is based on adversarial tribal groupings. Financing is based on corporate profit to the detriment of individuals, most of who live in poverty without basic human rights. The most important and influential world commercial bodies are the International Monetary Fund, the World Bank and the World Trade Organisation. Currently, none of their leaders are democratically elected. The leaders are placed there only with the final endorsement of the USA, and only if they pursue free trade and growth. One result of this has been the imposition of privatised water on Bolivia, which caused a revolt and the election of a new populist government who threw out the German water company.

Designers in all these fields have to rethink processes. They must redesign them to fit the demands of our time. Political theorists have to come up with better ways of world governance so that all the increasing populations can work together to share the decreasing resources. If we do not, we will be at war over those resources, such as water and oil. Economists have to design better economic models so that the whole world can share in prosperity. If we do not, millions will die of poverty and further millions of have-nots will fight back forcing the rich to defend themselves.

How can manufacturing supply the needs of everyone indefinitely, without using up all the materials and energy? The Western lifestyle is not about just material affluence – it also contains a spiritual richness gained from the beauty and vision of artists. How can we also ensure fulfilment of this need too? The solution lies in the creativity of designers.

As systems and processes are redesigned, so new needs for objects will arise. For instance, we should not be trying to design a new and better car, but a whole new transport system, starting with a completely open mind and no preconceptions. In fact, even before that, if transport is a problem, maybe we should be designing a better way of living that requires less transport? Only once the transport system has been formulated can we then look at what sort of vehicle might be required.

The same process can occur in ever smaller scales. I have run a number of design workshops on this subject, some of them at the Vitra Design Museum summer design school at Boisbuchet in France.[6] I ask the participants to rethink the ritual of eating. The first day is spent wholly in discussion, sitting in an intense circle on the warm French grass: how can we redesign eating in a better way? Is it just a bodily function or does it have any other functions? Two students from Mexico explained how food preparation in their country is a communal ritual, which provides the opportunity to discuss some of their social or moral issues. Without the ritual there would likely be less discussion. For us, how can the ritual be less wasteful? All waste has a value and maybe by changing the design of the ritual this value can be used? Where does the ritual actually begin?

Once the students have decided on the form of the ritual, then we start to think about what objects we will need. Now we are designing, not just to create something new for the sake of it, but because there is a real need. I insist that no new materials are bought in for this workshop, so that we use only what is left over from other workshops. We have to design solely around what is available. This is not a restriction, but a challenge and an opportunity to be creative and resourceful.

The workshop ends with everyone at Boisbuchet gathered around for the meal, which becomes a deeply meaningful event.

There are other ways in which design can help us meet the challenge of the future. Some designers use the object to convey a message. My favourite example of this is a drinking glass I was given in New York by the designer Nina Alesina. It simply has a hole drilled in it half way up the glass. Engraved on the glass next to the hole is a line with the words 'NEED' below, and 'WANT' above. If I fill it too full I get a wet shirt and I am reminded, is this something I really need or only want? So this object is not just a drinking tool but it also has the power to change, in the way that art does. And by taking up this responsibility, design is reverting to its original role as an aid to survival – not as a part of a threat to our survival.

This is where the hope for the future lies. Suddenly, designers find themselves in a critically important role. They have the training and the ability to question and re-evaluate every process, every system and every way of living and doing. No one in the developed world wants to give up more of their privileged lifestyle than he or she has to, or the comforts and luxuries that are taken for granted. So the challenge for designers is to find how to fulfil our needs in better, more sustainable and often different ways. The guiding principle has to be to fulfil our own needs without jeopardising the needs of future generations or those of poorer people in our world. If our grandchildren receive a legacy of depleted oil and water reserves, polluted and empty mine sites and rivers, and an overheated planet with hostile weather, they will sift through our landfills and ask, 'you did this for what?'

DESIGN CANNOT BE SEPARATED FROM CRAFT

Over recent years, the word 'craft' has come to have unwanted connotations in some quarters. A number of craft organisations have dropped the word 'craft' in favour of 'design' or 'form'. Craft is seen as obsessive and near-sighted, where craftspeople create singular and precious objects. Design on the other hand is perceived as idealistic and looking at a much larger canvas. But by marginalising the concept of craft, we are loosing aspects of making that provide an essential moral ingredient. Because embodied in hand making is the notion of care.

Design without craft is in danger of producing objects that have less intrinsic value. Mass-produced things are made quickly by machines to fulfil manufacturers' sales targets and they have no 'soul'. But the hand-made embodies the hand of the maker – it is made with care, with love even. This means that, compared to a mass-produced version, it will cost more, but it also will probably physically last longer, and the owner will be more attached to it and so be less able to freely throw it away. So, the hand-made product is a more sustainable product. They are qualities that can be added to objects to prevent the careless throw-away society of today.

But it goes beyond just an object itself, because I believe that the notion of care that craft represents is a vitally important metaphor for the whole of society. The indifference we manifest when we throw away paper cups and wooden chopsticks after just a moment's use extends to a general indifference towards the situation of other peoples and the planet itself. All that energy that was used to create and transport those items, both to your point of use and afterwards on to the landfill, was for that fleeting second of use. This happens because our society sees only short-term profit and neglects the greater picture.

If instead we had our own hand-made pocket cups and utensils that we cherished, we would both be caring for the planet and also promoting that paradigm or moral of care, which could be extended to other aspects of life. I am convinced that these are inextricably linked – if you care for the planet and people, you also care for small things, and vice versa.

The hand-made table that uses local wood from sustainably managed plantations is physically designed and built to last for generations. The distances from tree to maker to owner need only be a matter of kilometres. Ideally, the electricity required for making is also generated locally, without the line-loss waste of transit. With time, the timber will acquire a patina that will increase its attraction, and there will be stories and events embedded into the table's history. Within its lifetime, another tree will grow to replace the tree eventually if needed. This is sustainable design.

By contrast, a plastic table uses non-renewable, oil-based materials. The oil took millions of years to form from decaying forests. Once used, this valuable resource is gone forever. The table is one of thousands that are injection-moulded in minutes, largely untouched by human hand. It is made in a large central factory, powered by dirty coal-fired generators, probably in another country, and will be transported a long way to its final use. With time it will rapidly become tacky and devalued and will be quickly replaced by a newer more fashionable model because, by its very nature, there is much less reason to care for it. Can we really justify this one-off use and irretrievable loss of resource?

SUSTAINABILITY IS A STATE OF MIND
Either you care or you do not. If you care, there is no separation between things, people and the environment. Wetlands, rainforests, Africa's poor, future generations (your grandchildren) are all threatened by climate change, and if we care at all, we care about all of them. It is inbuilt in the psyche of the craftsperson to care about how he or she does things. As a woodworker, I preferred to spend more time trying to use up offcuts, than quickly cutting pieces out of a new plank.

This state of mind is repeated in every little action without thinking: paper is printed on both sides if it is printed at all, the tap is turned off for brushing teeth, plastic waste is separated from compost, car trips are combined or put off. It is not a question of rationale or thinking. Because if it were, you would wonder What is the point of such a miniscule gesture? Rather, you are doing it because you care and you cannot do otherwise.

Much of our current crisis is caused by the fact that there are too many people repeating seemingly miniscule wasteful actions too many times. It is the sum total which is the problem, not the individual actions. So, in the same way the problem can be solved by countless actions of care which in themselves may seem so irrelevant.

The situation we are now in calls for a new form of altruism that has no precedent in our history. People have fought wars for a better world but never in peacetime have people been asked to make sacrifices for which they will receive no personal reward. But if future generations, and less lucky people living now, are to benefit – or at least not be so disadvantaged – then we have to give up some of the luxuries of our lifestyle. There is no immediate financial, physical or social gain in this; only the reward of knowing you did the right thing. This requires an enormous moral shift.

It is in the nature of human beings to resist pressure. If told we must change, we will resent the pressure and resist. If people do not care, we cannot tell them to. If, however, we offer them a better alternative, they will choose to take it up themselves. This is the challenge to today's designers.

A DREAM

I see myself living in a small bach by the sea. The building is simply built, of materials that can be taken apart and used again many times. It relies on no outside services, generating all its own power and using all its own waste, which is not waste at all but a source of energy or nutrients. It is like a sailing boat pulled up on the shore – maybe I could rearrange its parts and launch it onto the sea one day? But I have all I need here within a few kilometres of my home.

In this beautiful place, my creative mind takes off on flights of fancy. I have redesigned the way that Aotearoa/New Zealand can still create its export earnings in a carbon-neutral world. I do my design work on a computer and put up the designs on my website. Anyone anywhere in the world can buy one, download it and customise it with colour, dimensions or texture how they wish. If it is small, they will create it once only on their home desktop 3D printer. Larger things will be created in a local computer controlled manufacturing plant. In either case the object is made using only local energy, and materials that can be replaced as fast as they are used. The buyer has their precious unique object to enrich his or her life with a little bit of my sunny beach, and I am able to earn my living without generating any carbon.

The bach is already half built ...

ENDNOTES
1   His lecture tour on climate change is documented in the controversial movie *An Inconvenient Truth*. 2006. Paramount Classics and Participant Productions.
2   Stern, N. 2007. Cambridge: Cambridge University Press. For more information, http://www.hm-treasury.gov.uk/Independent_Reviews/stern_review_economics_climate_change/sternreview_index.cfm (accessed 26 Sept 2007).
3   Wohlforth, C.P. 2004. *The Whale and the Supercomputer: On the Northern Front of Climate Change*. New York: North Point Press.
4   Walker-Morison, A. 2004. 'Materials Matter', *Architectural Review Australia 092*. St Kilda, Victoria: Niche Media.
5   These figures are taken from an Australian website: Trainer, F.E. (n.d.) The Simpler Way, which gives full references: http://socialwork.arts.unsw.edu.au/tsw/ (accessed 26 Sept 2007).
6   See www.boisbuchet.com (accessed 26 Sept 2007).

The role of ethics
in climate change strategies

Prue Taylor

Prue Taylor teaches environmental and planning law at the School of Architecture and Planning, University of Auckland. She is the Deputy Director of the New Zealand Centre for Environmental Law and a long-standing member of the World Conservation Union, Commission of Environmental Law and its Ethics Specialist Group. Her specialist interests are in the areas of climate change, human rights, biotechnology, environmental governance, ocean law and policy, and environmental ethics. Prue's book, *An Ecological Approach to International Law: Responding to Challenges of Climate Change*, won a Legal Research Foundation prize. In 2007, she received an Outstanding Achievement award from the World Conservation Union in recognition of her contribution, as a world pioneer on law, ethics and climate change.

THE CURRENT SITUATION
People are moral beings, but we have largely overlooked the role of ethics in climate change strategies. If we continue to do this, we will not achieve the necessary level of economic, cultural, social and spiritual transformation needed to avoid serious harm to the Earth's climate system. More comprehensive scientific information and new and more complex legal agreements will not, alone, be sufficient. Without a value-based foundation, science and law cannot dislodge our current exploitative approach to natural and social systems that disregard the Earth's finite resources.

THE VISION FOR 2020
There will be clear signs of a new ethical framework motivating individual and collective human behaviour. By 2020, people and organisations all over New Zealand will understand that everyone has responsibility for all peoples, all nations and the community of all life on Earth, now and into the future. This ethic of universal responsibility will be supported by specific principles that will guide all our decisions and actions. In the particular context of climate change, ethical imperatives will have replaced justifications for doing nothing or little, based on individual choice and self-interest. Further, our responses to climate change will be just one part of a widely accepted broader framework of sustainability, which will acknowledge that environmental protection, respect for human rights, equitable human development and democracy, non-violence and peace are interdependent and indivisible.

STRATEGIES TO ACHIEVE THE VISION
This vision can be achieved by widespread adoption of the *Earth Charter*. The *Earth Charter* can function as both a vision and a guide for the necessary ethical transformation. First and foremost the Charter should be used as the basis for full and frank ethical debate, in the course of which all values and ethical positions can be exposed, examined, discussed and reflected upon. Next, individuals, organisations, local authorities, churches, schools and business should develop their own set of guidelines based on a considered, ethical approach to environmental protection and social justice.

INTRODUCTION

Through the now famous movie, *An Inconvenient Truth*,[1] Al Gore sought to raise awareness around climate change. Much of the movie focused on presenting the scientific consensus on both causes and impacts.

However, in three separate instances he stated that climate change was a 'moral' issue. What did he mean by this? He did not explain. Perhaps his point required no elaboration because most viewers understood, at some level, what moral issues are and what moral action requires. (In fact, this turned out to be the case with a group of Auckland University students in the Planning Programme, but more on this later in the chapter). For now, ask yourself these questions: What does taking moral action involve? Does it mean that making changes and taking action is no longer a choice, but an obligation or responsibility? If so, to whom is this responsibility owed? How should you judge the benefits of your actions (for example, should you take actions that go beyond meeting your own self-interests)? What is the broader context within which action must be considered?

The purpose of this chapter is to provide an opportunity for exploring answers to the above questions. The aim is to inform you and thereby inspire you to take 'ethical action' in response to the threats of climate change. Ethical action is not just any action. As will be explained, it is action that is guided by: (1) an overarching ethical vision (or worldview) of humanity as a member of the community of all life on Earth and (2) a framework of ethical principles that enable, and guide, practical implementation of that vision.

WHY ETHICAL ACTION IS NECESSARY

It has long been recognised that human behaviour toward the Earth is leading to dreadful consequences. Further, it has long been understood that human behaviour will not change until and unless we realise that this behaviour is morally wrong and misconceived. It is behaviour that is based upon an unjust and false perception of our place within the natural world. Quite simply, we behave as if we were independent from the Earth's systems, and as a consequence we plunder its fruits faster than they can be replenished and pollute its reserves of air, water, sea and land with wastes faster than they can be absorbed. We do this regardless of our impact on ecosystems and at a rate and scale that are harming and endangering humans and non-humans. We behave in this manner to meet our own short-term wants and needs, irrespective of the interests of others. In short, this human-centred behaviour is a root cause of the Earth's degradation and changing it is critical. One of the key methods for doing this is by changing the worldview and value systems that drive this behaviour.

Perhaps the most famous statement of the need for, and about the source of, change in human behaviour was made in the 1940s when Aldo Leopold outlined what has become known as the 'land ethic'. It has continued to inspire and guide discussion ever since:[2]

> All ethics so far evolved rest upon a single premise: that the individual is a member of a community of interdependent parts. His instincts prompt him to compete for his place in the community, but his ethics prompt him to co-operate ... The land ethic simply enlarges the boundaries of the community to include soils, waters, plants, and animals or collectively: the land. In short, a land ethic changes the role of *Homo sapiens* from conqueror of the land-community to plain member and citizen of it. It implies respect for his fellow-members, and also respect for the community as such.

Leopold is reminding us here that humans have the capacity to develop ethics that can take us beyond our instinctive behaviours and responses.

What do we mean when we refer to problems as being moral or ethical issues? We mean that the consequences of our actions have become such (for example, so harmful to others and to community or society) that we need to develop an ethic and framework for understanding and judging what is 'right' or 'wrong', 'good' or 'bad'. Historically, our religious and spiritual traditions have provided societies with ethical frameworks. However, in more modern times, when the influence of religion and spirituality has diminished and new problems arise, then societies must themselves develop new ethical frameworks. This is what happened following the horrors of World War II. World leaders came together to try and ensure that these horrors would never be repeated and, from their work, we benefit from the legal protection of human rights and legal prohibition of genocide, war and the use of nuclear weapons (to name a few).

Unfortunately, the writers of the 1945 United Nations Charter did not address the future of planet Earth. However, this does not mean that people were unaware of emerging problems. In 1948, the famous writer Aldous Huxley wrote a stirring letter to his brother. This brother went on to be a founder of the world's oldest and largest environmental body, the World Conservation Union. Huxley wrote:[3]

Meanwhile I come to feel more and more that no system of morals is adequate which does not include within the sphere of moral relationships, not only other human beings, but animals, plants and even things. We have done quite monstrously badly by the Earth we live in, and now the Earth we live in, with its soil eroded, its forests ravaged, its rivers polluted, its mineral resources reduced, is doing so badly by us that, unless we stop our insane fiddling at power politics and use all available knowledge, intelligence and good will to repair the harm we have done, the whole of mankind will be starving in a dust bowl within a century or two. People still seem to believe that there is poverty in the midst of plenty, when in fact there is only poverty – and all through our own fault, through not treating nature morally ...

If we do not do something about it pretty soon, we shall find that, even if we escape atomic warfare, we shall destroy our civilisations by destroying the global capital on which we live.

Given this powerful message, it is perhaps no surprise that from its inception, in 1948, the World Conservation Union has exercised moral leadership. It has consistently promoted understanding and legal recognition of the need for ethical action. Two of its most important initiatives were the 1980 *World Conservation Strategy* and the 1982 *World Charter for Nature*. The former stated:[4]

> Ultimately the behavior of entire societies towards the biosphere must be transformed if the achievement of conservation objectives is to be assured. A new ethic, embracing plants, animals as well as people, which will enable human societies to live in harmony with the natural world on which they depend for survival and well-being.

And the *World Charter for Nature*, which was adopted by the United Nations General Assembly, articulated a new worldview and a range of ethical principles including the statement that:[5] 'every form of life is unique, warranting respect regardless of its worth to man, and, to accord other organisms such recognition, man must be guided by a moral code of action.'

The efforts of the World Conservation Union did not, alone, create the necessary momentum for change. However, they were critical to a new generation of effort. In 1992, spurred on by the failures of the United Nations Conference of Environment and Development, global civil society began the task of developing a new ethical vision and series of principles to correct and fundamentally transform human behaviour. Foremost among these efforts is a document known as the *Earth Charter* (see later in the chapter for the story of the *Earth Charter* and its relevance to the objective of achieving carbon neutrality in New Zealand). However, a relevant question at this point is: Why is scientific knowledge alone not sufficient to motivate fundamental human behavioural changes? Or, to put it another way, once we understand the science better, will we not behave differently?

In exploring this question, it is important to realise that we stand at a defining point in human history. Our 'scientific' understanding of our relationship with Earth's systems has never been greater. This is best illustrated by the 2005 *United Nations Millennium Ecosystem Assessment Report*.[6] This report was designed as a comprehensive and authoritative evaluation of human impact on ecological systems. It makes many important conclusions, but for our purposes one is critical. Despite some 50 years of scientific endeavour, together with considerable international legal activity to manage ecological degradation, more ecological harm has occurred in the last 50 years than at any time in recorded human history.[7] This conclusion establishes that we are not responding adequately to scientific knowledge, or accepting the implications of this knowledge. Our responses (such as they are) are not going far enough or deep enough. Nor do they address fundamental structural issues such as the current pursuit of economic growth. But scientific evaluations do not tell us why we are not responding. Is there a plausible answer to this apparent paradox? Could it be that humanity is indulging in 'denial'; that is, consciously and unconsciously refusing to acknowledge that a human-centred and exploitative value system is no longer viable in a finite world, despite abundant scientific evidence that this is the situation? Heads in the sand while the tide rises?

If humanity is in denial about the harm wrought by an exploitative value system, and therefore the need to change it, then for change to occur something very fundamental will be needed. Of course, massive natural disasters on the scale of the droughts and hurricanes recently experienced in developed nations such as Australia and the USA can be the motivators for change. But they are not the only possible motivators and waiting for them to come along is nothing short of irresponsible. Rather, 'major social transformation' needs to happen now. Changes of this nature and magnitude do not occur by the power of scientific evidence alone, but also involve a change in individual and group ethical values.

At the level of international politics, Maurice Strong has been one of the most consistent voices of the view that better scientific, legal or economic resources will not take humanity far enough, fast enough. In his words:[8]

> Economic change is imperative, indeed critical. But, in the final analysis, economic factors, like other aspects of human behavior, are deeply rooted in the human, cultural, spiritual, social and ethical values which are the fundamental sources of motivation of the behavior of people and nations. Technocratic measures can facilitate but not motivate solutions to basic issues … The practical solutions we devise, the concrete measures we propose will be of little effect if they are not accompanied by a deep and profound stirring of the human spirit.

Responses that stop short of a transformation in the prevailing ethical framework will be insufficient because they will only address symptoms, but perpetuate the cause. It may be difficult to believe the necessary changes of heart (or spirit) and mind are possible. This is where the *Earth Charter* becomes critical.

THE *EARTH CHARTER*: A VISION AND A GUIDE

The *Earth Charter* (see Appendix 2) is a document that seeks to transform human understanding and behaviour, for the purpose of creating a sustainable, just and peaceful global society. It sets out a widely recognised ethical vision and framework and builds upon the efforts of the World Conservation Union and others. But to fully understand its purpose and significance, we need to retrace the history of its creation and the evolution of its status.

In 1987, the World Commission on Environment and Development recognised the need for a new ethic to underpin and guide sustainable development. Accordingly, a charter, intended to provide an ethical foundation for other international legal agreements, was negotiated in the run up to the 1992 Earth Summit, a United Nations conference on the environment and sustainable development. Unfortunately, this draft charter failed to attract sufficient support from United Nations member states. However, representatives of non-government organisations (for example, environmental and social groups) were more successful. During the Earth Summit they succeeded in producing the first draft of a document that became known as the *Earth Charter*. This draft became the starting point for a worldwide consultation and drafting process, involving thousands of individuals and organisations from around the world, and lasting some seven years. In 2000, a final draft of the *Earth Charter* was adopted and launched. Its status as the first global civil society document of its kind is directly attributable to the process by which it was created.

The very inclusive and participatory process used to create the *Earth Charter* is also crucial to its purpose; the creation of an inclusive – or shared – ethical vision of the values and principles needed to guide humanity. It is concerned with the identification and promotion of ethical values that are widely shared by all nations, cultures and religions. Thus, its drafters aimed to create a document that was representative of the current and ancient wisdom and knowledge of as many of the world's diverse peoples as possible.

The essence of the *Earth Charter* can perhaps best be understood by reading the first four core principles, upon which all the other supporting principles are based:[9]

1. Respect Earth and life in all its diversity;
2. Care for the community of life with understanding, compassion, and love;
3. Build democratic societies that are just, participatory, sustainable, and peaceful;
4. Secure Earth's bounty and beauty for present and future generations.

These four principles create a new ethical vision that is intended to form the basis of a very different relationship between humanity and nature. It is an ethical approach that gives rise to universal responsibility to actively care for and respect the community of all life. This means that responsibility is owed by each and every human being, however we conduct and organise ourselves, either as individuals, communities, organisations, businesses, governments or transnational organisations. Universal responsibility, therefore, acknowledges not only the interconnected local and global dimensions of our problems, but also that each of us has a vital role to play. The challenges will be great and creative leadership will be pivotal to implementing this responsibility. But so will the aggregate efforts of individuals.

What then do 'care' and 'respect' mean in the context of the *Earth Charter*? Accepting responsibility to care for the community of life involves taking into account what ecosystems need to be able to maintain their own health and to regenerate so that they can continuously provide the basis for all life. Care also involves ensuring that fellow humans have the opportunities necessary to participate in the community of life. Treating Earth with respect assumes a positive or deferential appreciation of ecosystems as the basis of all life. Respect also entails recognition that ecosystems and all other beings, human and non-human, have intrinsic value (meaning value in their own right, irrespective of worth to others). In respecting, we accept that short-term sacrifices may be required in order for long-term benefits to be realised.

Now, let us return to the questions raised at the start of this chapter to see what answers the *Earth Charter* provides. They were: What does taking moral action involve? Does it mean that making changes and taking action is no longer a choice, but a responsibility? If so, to whom is this responsibility owed? How should you judge the benefits of your actions? Should you go beyond self-interest?

In response to the first two questions, the *Earth Charter* provides clear answers; moral action is based upon an ethic of care and respect for the community of all life, and creates a universal responsibility to take action. In response to the third question, the responsibility is owed to the community of all life, that is humans and non-humans, and present and future generations. And it is also clear that we must go beyond self-interest. In respecting, we impose limits upon ourselves. Given that so many aspects of our current lifestyles are achieved by exploiting both the Earth's resources and the human labour forces of poor nations, taking appropriate action will require going beyond the limited measure of 'self-interest' and understanding the larger 'community of interest'.

The *Earth Charter* has two particularly noteworthy strengths. First, ethical visions often fall short of having the impact we would like them to have because they are expressed at a level of generality that makes them hard to apply. To avoid this, the *Earth Charter* sets out 12 specific practical principles, all of which are consistent with the vision. Within these 12 specific principles are three to five sub-principles that provide even further practical detail. So, for example, Principle 5 states that to fulfil the vision it is necessary to:

> Principle 5: Protect and restore the integrity of Earth's ecological systems, with special concern for biological diversity and the natural processes that sustain life.

Its sub-principles include:

- Adopt at all levels sustainable development plans and regulations that make environmental conservation and rehabilitation integral to all development initiatives;
- Promote the recovery of endangered species and ecosystems;
- Manage the use of renewable resources such as water, soil, forest products, and marine life in ways that do not exceed rates of regeneration and that protect the health of ecosystems;
- Manage the extraction and use of non-renewable resources such as mineral and fossil fuels in ways that minimise depletion and cause no serious environmental damage.

In this manner, the *Earth Charter* provides us with practical ethical principles that we can use to both guide and assess our actions.

The second special strength of the *Earth Charter* is that it is not exclusively concerned with the environment and activities that are of immediate harm to the environment. It does have a special emphasis on the environment, but its inclusive ethical vision recognises that environmental protection, respect for human rights, equitable human development and democracy, non-violence and peace are interdependent and indivisible. Thus, it creates a new integrated framework for thinking about and addressing all these issues. The result is a broad conception about what constitutes sustainable community and sustainable development. Again, some examples help illustrate this point. The burning of fossil fuels and the felling of forests are important causes of climate change. In many poor parts of the Earth, extreme poverty means that people have no choice but to cut trees to use as firewood, for cooking and warmth, or to sell. To stop this happening, we will need to find ways to address human poverty.

A further example is provided by biofuels; they must not be produced in a manner that deprives people of diverse and productive use of their land or that creates further hunger by reducing grain crops in favour of biofuel feed stocks (such as corn and sugar cane). For these reasons, the *Earth Charter* includes principles on social and economic justice. Principle 10, for example, states that it is necessary to: 'ensure that economic activities and institutions at all levels promote human development in an equitable and sustainable manner.' Principles 10a and 10c state the need to: 'promote the equitable distribution of wealth within nations and among nations' and 'ensure that all trade supports sustainable resource use, environmental protection, and progressive labor standards.' These principles attempt to address the fact that much of the wealth and well-being enjoyed in wealthy nations is achieved at the expense of the people and environment in poor nations. They, together with other *Earth Charter* principles, remind us that reduction of New Zealand's own greenhouse gas emissions is only one aspect of a multi-faceted issue. A refusal to buy unsustainably produced products, or goods produced by exploited labour forces, combined with support for increased financial and technical aid for developing nations, are important and related actions to take.

The *Earth Charter* is intended to be transformational and, as such, it challenges much conventional thinking and practice. Gaining worldwide acceptance of the *Earth Charter* will not be easy; indeed, there has already been resistance to it from powerful nations (such as the USA) that have a vested interest in the status quo.[10] However, the hope is that the *Earth Charter* will evolve in a similar manner to the Universal Declaration on Human Rights. This declaration evolved over time to gain international legal status and to become the basis for many international treaties on human rights. There are already early indications that the *Earth Charter* is well supported both by civil society and by a growing number of 'enlightened' nation states. To date, it has been formally endorsed by over 2,000 organisations worldwide.[11] Two significant recent international endorsements came from the United Nations Educational, Scientific and Cultural Organization (UNESCO)[12] and the World Conservation Union that comprises some 77 nation states, 114 government agencies, and 800 NGO members.[13] The UNESCO endorsement is important because it adopts the *Earth Charter* as the ethical basis for the United Nations Decade of Education for Sustainable Development. Another important endorsement is from the International Council for Local Environmental Initiatives (ICLEI). ICLEI has approximately 500 local government members, all committed to an 80 per cent reduction in GHG emissions by 2050.[14] In New Zealand, at central government level, the Parliamentary Commissioner for the Environment has endorsed the *Earth Charter* and the Prime Minister has unofficially acknowledged it.

Endorsement of the *Earth Charter*, at individual, group, organisational and state levels are all important, but the future of the *Earth Charter* will depend upon its use as a tool in the creation and guidance of dialogue. The reason for this is both simple and important; new ethical frameworks cannot be imposed upon societies. They must grow out of global dialogue, undertaken at all levels. The historical merit of the *Earth Charter* is that it has initiated this dialogue, both during its drafting and since its adoption in 2000. The ongoing task is for both advocates and opponents of it to use the *Earth Charter* as a basis for full and frank ethical debate, in the course of which all values and ethical positions are exposed, examined, discussed and reflected upon. It is only once this occurs that fully reasoned and considered choices, about issues like climate change, can be made.

PRACTICAL SUGGESTIONS FOR USING THE *EARTH CHARTER*

The practical suggestions below have been selected to fit the New Zealand context, focus on climate change (but not artificially separate it from its proper context) and cover a range of levels – individual, community, business, and local and central government. Actual examples of many of the actions summarised below, can be found in the inspirational book: *Toward a Sustainable World: the Earth Charter in Action*.[15] The hope is that the suggestions described below will stimulate a range of efforts to respond to the global and local challenges posed by climate change.

The *Earth Charter* can be used as:
A guide for internal reflection and consideration of fundamental ethical values and attitudes.

From awareness comes appropriate ethical action. This internal reflection can lead to explicit dialogue within families, schools, universities, communities, work places and business organisations. Once this internal reflection and dialogue occurs, the *Earth Charter* can then be used as the basis for the creation of an 'ethical code of conduct' (professional, institutional and personal) to guide and then assess decision-making and action.

In simple terms, this can help move people beyond inertia, by making them aware that there is no longer a choice about whether or not to take action in response to climate change, but that there is now a universal ethical responsibility to do so. One simple way of making this very real is to endorse the *Earth Charter*. This involves a pledge to promote the *Earth Charter*, to contribute to action projects and to implement its principles, by applying it to your organisation or personal life, whichever is appropriate.[16] Endorsement could (but need not) lead to involvement in the global dialogue on addressing climate change. This is a global online dialogue focusing on responding to climate change from an integrated ethical perspective.[17] In particular, this dialogue emphasises the role of an ethic of universal responsibility in motivating people to go beyond self-interest and nations to go beyond national interest. Both are essential if we are to achieve the necessary depth and scale of change.[18]

A map of the interconnectedness of the challenges facing humanity
Identifying the connections of our world – global to local and local to global – is helpful, particularly in combination with mapping the interdependent dimensions (environmental, social, economic, cultural and political). This can assist a business, for example, to audit all of its operations in a comprehensive manner. It would be counterproductive to upgrade a vehicle fleet to reduce greenhouse gas emissions while continuing to sell goods and services that exploit vulnerable ecosystems and peoples. At an individual level, we need to see the connections between fitting energy efficient lightbulbs in our homes and buying cheap imported products that are made in environmentally harmful ways and involve exploitation of poor workers. Both are critically important because climate change is about more than just domestic greenhouse gas reductions.

A check list for ethical spending and ethical investment As consumers (or purchasers of supply) and investors (individually or through managed funds) we have enormous collective power. Clearly, buying products made from tropical rainforest timber should be avoided. Not only does this encourage destruction of livelihoods, habitat and ecosystems, but the rainforest removal means that we have lost what are considered to be the most effective sinks for CO_2. Even if we buy products and services from companies that claim to have off-set greenhouse gas emissions, we need to check if those off-sets are achieved by setting aside forests, in poor nations, in a manner that deprives the local people of a source of food and income. This is epitomised by the question: Whose forest? As regards ethical investment, investing in corporations that are not taking significant responsive action, or are themselves engaged in supporting harmful activities, should be avoided.[19] The *Earth Charter* can be used, in a related way, to give force to shareholders resolutions and submissions, ensuring ethical investment of taxpayer's money via superannuation investments.

An educational tool The *Earth Charter* has and is being extensively used within primary and secondary schools and tertiary institutions to guide learning about the critical choices facing humanity, the scope of change required and the role of ethics. It can be used to complement existing government educational efforts such as Enviroschools (see Chapter 3) and the recent national curriculum review. This curriculum review seeks to include teaching of ecological values as part our national school curriculum. The *Earth Charter* itself is a valuable tool for the teaching of ecological values, because it includes (but goes beyond) respect for human needs. People often wrongly assume that ecological values deny human needs. Practical use of the *Earth Charter* is most developed in the area of education, partly as a result of the UNESCO adoption mentioned above.

The Earth Charter Initiative[20] has developed an extensive website that includes resources, projects, activities, workshops and sample curricula. All this material is of practical use, demonstrating how the *Earth Charter* can be used as an essential component of education for sustainable development.[21] A related practical application is through the Earth Charter Youth Initiative. This takes the form of an action-orientated youth network facilitating direct action locally, nationally and internationally.[22]

The basis for cross-cultural and interfaith dialogue in society The *Earth Charter* can be used to foster understanding between diverse cultural groups in New Zealand. As Charter Principle 12 states, it is necessary to: 'uphold the right of all, without discrimination, to a natural and social environment supportive of human dignity, bodily health, and spiritual well-being, with special attention to indigenous peoples and minorities.' In the New Zealand context, we need to ensure that our climate change responses do not cause further harm to indigenous and minority peoples. Failure to do so could result in loss of human dignity and spiritual and cultural well-being, creating a negative flow on effect. Again, the Earth Charter Initiative can be of practical help. It is currently developing a new programme on religion and sustainability. The objective is to support further engagement from the world's spiritual and religious communities, acknowledging their vital current and future role in encouraging changed values and behaviour.[23]

The basis for the development of a community sustainability plan
This can be achieved by use of the Earth Charter Community Action Tool (EarthCAT), which is a guide for the creation and implementation of such a plan. This tool enables communities to learn from the experience of others as well as monitor progress via implementation indicators such as those dealing with multiple facets of climate change mitigation and adaptation.

CONCLUSION
The introduction to this chapter referred to University of Auckland students, participating in the Planning Programme, and to an understanding of what moral action requires. This story demonstrates the powerful role the *Earth Charter* can play.

The Planning Programme includes a short module on environmental ethics, and this includes an introduction to and discussion on the *Earth Charter*. The objective is to discuss values and behaviour and viable alternatives to our current exploitative value system. In 2007, a new planning course was offered entitled 'Climate Change and Planning'. All of the students in this course had completed the module on environmental ethics. The climate change course began with the students viewing and analysing *An Inconvenient Truth*. This analysis involved class discussion based upon the integrated approach of the *Earth Charter*, linking social, economic, cultural, spiritual and ethical causes of climate change. When reading the written evaluations, I was struck by the number of students who referred to the moral assertion made by Al Gore and then linked this to the class discussion based on the *Earth Charter*. The result was a much deeper and interconnected understanding of the changes needed and how this could be translated to motivate individual actions. As one student noted (and many echoed), they now understood that taking constructive action was no longer a matter of individual choice but a moral responsibility, owed to current and future generations of humans and other living beings both in New Zealand and in other parts of the world. Further, and of no less importance, was the understanding that actions should no longer be justified solely by criteria of self-interest as this would limit the scale of actions taken. For example, reduced electricity consumption leads to saving money on power bills. Similar savings could not always be expected as a result of using public transport or purchasing locally produced goods. The extra costs associated with these actions were not used to justify inaction, but were accepted as the current price of acting responsibility. Such shifts in the behaviour of students may seem small, but when multiplied (in whole or in part) across a class of 55 students, the cumulative impact is both exciting and dramatic. Of course, the real challenge is to make value and behavioural changes enduring. Even taking this into account, I am convinced that the *Earth Charter*, and *An Inconvenient Truth*, came together to mutually reinforce one another and create a powerful impetus for ethical action on climate change. This convergence certainly provided me with one of my most rewarding teaching (and learning) experiences.

ENDNOTES

1 *An Inconvenient Truth*. 2006. Paramount Classics and Participant Productions.
2 Quoted in Robinson, N. 2002. The 'ascent of man': legal systems and the discovery of an environmental ethic. In Craig, D., Robinson, N. and Kheng-Lian, K. (Eds). *Capacity Building for Environmental Law in the Asian and Pacific Region: Approaches and Resources*, Volume 1 (pp. 104–105), Manila: Asian Development Bank.
3 Quoted in 'Shaping a sustainable future: The IUCN Programme 2009–2012', p. 6 (to be adopted at The World Conservation Congress, Barcelona, Spain, 5–14 October 2008).
4 IUCN. 1980. *World Conservation Strategy – Living Resource Conservation for Sustainable Development*. Section 13.1. IUCN: Gland, Switzerland.
5 World Charter for Nature, 22 International Legal Materials, 455 (1983), Preamble.
6 The *United Nations Millennium Ecosystem Assessment Report* is available at: www.MAweb.org (accessed 27 Aug 2007).
7 *United Nations Millennium Ecosystem Assessment: Ecosystems and Human Well-being: Synthesis* (p. 26). Available at: www.maweb.org/en/index.aspx.
8 Quoted in Hassan, P. 2005. Earth Charter: An ethical lodestar and moral force. In Corcoran, P., Vilela, M., & Roerink (Eds.). The Earth Charter in Action: Toward a Sustainable World. (pp. 29–30). Amsterdam: KIT Publishers.
9 A copy of the *Earth Charter* is available at: http://www.earthcharter.org (accessed 03

June 2007). The structure of the *Earth Charter* is significant. It is described by Mackey as follows: 'Structurally, the Earth Charter comprises an introduction, ("Preamble"), a set of principles written in the style of ethical imperatives, and a concluding statement ("The Way Forward"). There are 77 principles organised around four main themes: (1) Respect and Care for the Community of Life; (2) Ecological Integrity; (3) Social and Economic Justice; and (4) Democracy, Non-violence and Peace. Each theme has four main principles, each of which has a varying number of supporting principles. The principles in the first theme identify four core ethical commitments. Following these is a linking sentence: "In order to fulfill these four broad commitments, it is necessary to …" This makes it clear that the 65 principles under the following three themes are action-oriented in that (if accepted) they impose an obligation for people to give them due moral consideration and modify their behaviour accordingly to advance the first four core ethical commitments.' Mackey, B. 2004. The Earth Charter and ecological integrity – some policy implications. *Worldviews* 8, 76–92.
10   IUCN World Congress (in respect of Congress Resolution WCC3.022) that endorsed and recognised the *Earth Charter*. The USA tabled a written objection to adoption of this resolution.
11   See www.earthcharter.org for information on endorsements (accessed 27 Aug 2007).
12   UNESCO adopted a resolution in 2003 recognising the *Earth Charter* as an important ethical framework for sustainable development and an important educational instrument. See www.unesco.org/education/ (accessed 03 June 2007).
13   At the 2004 World Congress, the IUCN adopted resolution WCC3.022, endorsing the *Earth Charter* as an inspirational expression of civil society's vision for building a just, sustainable and peaceful world. It also recognised the Earth Charter as an ethical guide for future IUCN policy.
14   www.iclei.org/index.php?id=800 (accessed 27 Aug 2007). Twenty-seven New Zealand local authorities participate in the ICLEI Communities for Climate Protection Programme: www.iclei.org/index.php?id=3931 (accessed 27 Aug 2007).
15   Holgate, M.W. 2005. *Toward a Sustainable World: the Earth Charter in Action.* Amsterdam: KIT Publishers.
16   For information on endorsement, see www.earthcharter.org (accessed 03 June 2007).
17   www.earthcharterinaction.org/climate/ (accessed 27 Aug 2007).
18   Mackey, B. and Li, S. Winning the struggle against global warming – what will it take? Available at www.earthcharterinaction.org/climate/ (accessed 27 Aug 2007).
19   For example, RepuTex is an Australian agency that assigns ratings to companies, organisations, and government bodies based on their social, environmental, corporate governance, and workplace practices. It uses *Earth Charter* principles for the creation of its benchmarks. Its work provides an innovative way for the *Earth Charter* to be useful in the business sector, www.reputex.com.au/ (accessed 27 Aug 2007).
20   The Earth Charter Initiative is the collective name for the global network of people, organisations, and institutions who participate in promotion and practical implementation of *Earth Charter* principles. For more information, see www.earthcharter.org (accessed 27 Aug 2007).
21   See www.earthcharter.org (accessed 27 Aug 2007).
22   See www.earthcharter.org (accessed 27 Aug 2007).
23   See www.earthcharter.org (accessed 27 Aug 2007). See also Gardner, G., 2006. *Inspiring Progress: Religions' Contributions to Sustainable Development.* London: W.W. Norton & Company.

Responsible investing

Rodger Spiller

Rodger has carried out research in responsible investment for over 20 years, and holds a PhD in business ethics, socially responsible investment and responsible business. He is Managing Director of Money Matters (NZ) Ltd and provides responsible investment advice for individuals and organisations. His consultancy business, Rodger Spiller & Associates, undertakes research and trains and guides companies to improve their environmental, social and economic performance. He is New Zealand Director of the Responsible Investment Association and a Director of Transparency International (NZ). Rodger is a former member of the Securities Commission.

THE CURRENT SITUATION
Most of today's investors are not choosing to invest responsibly and so are missing out on the opportunity to make a difference as well as making money. One of the reasons why people do not invest responsibly is that they perceive incorrectly that they cannot make money and address environmental challenges, like achieving carbon neutrality, and social challenges, such as seeking solutions to poverty. Another reason is that many people are not aware of the options that exist or do not know how to invest responsibly.

THE VISION FOR 2020
By 2020, responsible investing will be the norm in New Zealand. Most New Zealanders will have a responsible investment plan and portfolio, including a KiwiSaver scheme. All investment advisers will have been trained in responsible investment and be able to converse intelligently with their clients on social and environmental matters. Their newsletters and seminars will feature examples of investments that help create a sustainable environment, economy and society. All fund managers will use responsible investment criteria as part and parcel of their mainstream investment activity.

Through applying the principles of responsible investment and business, New Zealand will be a world leader. The New Zealand Superannuation Fund and other crown financial institutions will have helped create a sustainable country for its citizens. Business will be regarded by the community as a powerful force for good. Maori businesses will be flourishing and demonstrating the power of adopting a quadruple bottom line of optimising environmental, cultural, social and economic wealth over generations. There will be ample funding for initiatives such as environmental projects, sustainable agriculture initiatives, holistic healthcare and education. All schools will teach responsible business and investment as part of their core curriculum.

Responsible investment and responsible business will be mainstream, with investors and businesses doing well and doing good.

STRATEGIES TO ACHIEVE THE VISION
First and foremost, individual New Zealanders need to to invest responsibly. We need to choose financial advisers and fund managers that offer responsible investment. Those already investing responsibly need to encourage others to do the same. All investors need to use their economic vote and financial advisers and fund managers will respond to market demand. Obviously, investors need to know whether a fund manager offers responsible

> investment. The Responsible Investment Association's SRI symbol testifies to the consumer that a fund systematically addresses environmental, social or ethical considerations. It also assures investors that a financial adviser has completed training in responsible investment and offers responsible investment advice. Disclosure requirements are a low-cost and efficient way of achieving this and the government intends to make it mandatory from 1 April 2008 for KiwiSaver fund managers to disclose their approach to Responsible Investment. Other, similar policies and regulations are needed.
>
> In a very real sense, the rate of progress towards a carbon-neutral 2020 will, to a large degree, be determined by the amount of investment in responsible businesses.

INTRODUCTION

We all know that we have to do something with our money, including accumulating enough of it to support ourselves in our old age. Some may also want to provide financial support for family and others. We have all invested, directly or indirectly, in investment funds and these are now the world's largest single owners of company shares. These funds are collectives of individuals like you. This means that when you put your money in an investment, you become part of a global conversation about which businesses are worth supporting. Given the enormous influence of business on human life and the impact we are having on this planet, we should be thinking carefully about what our money is saying. We can all use our money to encourage 'responsible investment' (also known as 'socially responsible investment' or 'ethical investment'), adding to a collective force that will ultimately lead to more and more businesses acting responsibly.

Responsible investment takes account of environmental and social as well as financial performance, and encourages companies to address the world's environmental and social challenges as business opportunities. The goal of carbon neutrality is seen as good corporate social responsibility and a growing list of companies are announcing dates for when they intend to become carbon neutral, often having been influenced by the demands of responsible investors. Responsible investment can also direct money away from companies doing harm, thereby reducing the negative effects of irresponsible business on the environment and society.

Many people, however, believe that making money and being true to your principles do not go together, and this perception stops the majority of people from investing responsibly. This in turn limits the money available for financing an environmentally and socially responsible world. Without the money required, we will not have a carbon neutral and sustainable society. The idea that you can have profit or principle, make money or make a difference, or that there needs to a tradeoff between doing well and doing good, is a fallacy. The tyranny of 'or' needs to be replaced with the wisdom of 'and'. We need to move from focusing only on making money or, alternatively, only on making a difference. This can provide a better environment, greater wellbeing and happiness for the individual investor, better businesses, and a better quality of life for the community as a whole.

RESPONSIBLE INVESTORS DO WELL
In my role as a financial adviser I am often asked by investors whether they can make money from responsible investment. The short answer is yes, and the figures presented later in this section speak for themselves. Investors benefit by avoiding the high costs of irresponsible business and reaping the benefits that tend to accrue to responsible businesses.

The high costs of irresponsible business
There are many well-known examples of irresponsible and unethical business practices. The famous Ford Pinto case in the USA is a dramatic illustration, and it became a turning point in public opinion about business responsibility. In 1978, three young women were burned to death when the Pinto they were driving was struck from the rear and the fuel tank exploded. The Ford Motor Company was subsequently indicted and tried on charges of criminal homicide – a first in corporate history. In his instructions to the jury, the judge said that Ford should be convicted if it could be shown that it had engaged in 'plain, conscious and unjustifiable disregard of harm that might result (from its actions) and the disregard involves a substantial deviation from acceptable standards of conduct'. Ford's legal representatives argued that the Pinto had met all applicable safety laws and regulations and was comparable to other cars of the same type and that since the Pinto was in compliance with the law, nothing further could be required of Ford. In 1980, the jury returned a verdict of not guilty.

It later became known that Ford, knowing the tank was vulnerable to rear-end collisions, had conducted an analysis comparing legal costs they might incur from burn deaths and injuries versus the costs of making the Pinto fuel tank safer. The study concluded that it would be less costly for the company to pay for deaths and injuries even though the safety changes needed would cost only $11 per car. Their calculations were sent awry when, even after being cleared of criminal charges in a court of law, they lost in the court of public opinion and subsequently paid out millions of dollars in civil suits.

Another prominent example of irresponsible business practices was Union Carbide's disaster in Bhopal in India in 1984. In the heart of the city of Bhopal, a Union Carbide subsidiary pesticide plant released 40 tonnes of methyl isocyanate (MIC) gas, immediately killing nearly 3000 people and, it is claimed, ultimately causing at least 20,000 deaths. Bhopal is frequently cited as the world's worst industrial disaster. Critics argue that around 500,000 people were exposed to the leaks; 120,000 continue to suffer from after-effects such as breathing difficulties, cancer, serious birth defects and blindness; and 50,000 people are unable to work. They maintain that a contributing cause to the explosion was the location of the plant. Reportedly, authorities had tried and failed to persuade Carbide to build the plant away from densely populated areas, and Union Carbide explained its refusal in terms of the expense such a move would incur. It was also claimed that Union Carbide previously produced their pesticide without MIC, but after 1979 began using MIC because it was cheaper – even though other manufacturers did not. The factory was reportedly operating at a loss and over-producing MIC that was not being sold, leading to a series of cost-cutting measures affecting the workers and their conditions. Clearly Union Carbide's unethical attempt to save money was very costly in the end.

The case of Exxon's response to the *Exxon Valdez* oil spill, the largest in USA history, is a high profile example of environmental irresponsibility. When the ship struck Bligh Reef in Prince William Sound, Alaska, on 24 March 1989, there was a spillage of more than 11 million gallons of crude oil, with enormous consequences for the marine habitat. Mass mortality of the wildlife included unprecedented numbers of seabird deaths – estimated at 250,000 in the days immediately after the event – as well as 1000-1800 sea otters and over 300 harbour seals. The spill cost around US$7 billion, including the clean-up costs. US$5 billion of this was the largest punitive fine ever handed out to a company for corporate irresponsibility. Exxon failed to show that it had effective systems in place to deal with a crisis and showed little leadership after the event. John Devens, the Mayor of Valdez, commented that the community felt betrayed by Exxon's inadequate response to the crisis, in contrast to the promises they had given of how they would react in such an eventuality. Exxon lost market share and slipped from being the largest oil company in the world to the third largest.

More recently, the corporate governance scandals and collapses involving Enron, WorldCom and Parmalat have put the spotlight on business ethics. All around the world these and other cases have highlighted how shoddy business ethics puts investors' money at risk.

How responsible businesses do well

In contrast to the above examples, responsible businesses do well financially by winning the support of stakeholders. These stakeholders include customers, employees, the community and shareholders.

Customers Customers are increasingly choosing to purchase from companies that have a reputation for responsibility. Across the EU, for example, one in six consumers either buys or boycotts a product because of the reputation of the company.[1] Here in New Zealand, 19 per cent of consumers are making purchasing decisions each month based on green or social imperatives.[2] Many customers are attracted by responsible business and no one is likely to boycott a business because of its responsible approach. So, responsible businesses can reach a greater share of the market, which in turn makes them more attractive investments.

Employees In an environment of low unemployment, being an employer of choice is a more valuable asset than ever. Responsible businesses are better able to attract and retain staff. This is particularly the case with Generation Y employees, because many young people are keenly interested in business responsibility. In a study of 11 leading European and USA business schools, 97 per cent of MBA graduates said they were willing to forgo financial benefits (on average 14 per cent of expected income) to work with an organisation with a better reputation for corporate social responsibility and ethics.[1] Reducing barriers to attracting the best and the brightest, and keeping them on board means greater productivity and performance and, in turn, increased profits for shareholders.

The community The community at large is also an important driver of financial performance. Business needs supportive relationships with the community in order to survive and thrive. The community provides businesses with a social licence to operate, and those businesses that act responsibly and build goodwill will be less prone to regulation and constraints on their entrepreneurial ability and the financial returns to shareholders that this generates. By strengthening community relations, business benefits from the improved wellbeing of people in the community, some of whom are prospective or existing employees, customers, suppliers and shareholders. Studies of the community's perception of businesses that it most admires show that these businesses are usually those with the highest level of social responsibility.[1]

Shareholders Shareholders benefit financially through this support from other stakeholders. The risks are reduced and returns increased, making shareholders more likely to invest and retain their shares in companies that are responsible. This in turn increases share prices and the capital available for responsible business. Responsible investment and business practices, therefore, are actually a manifestation of enlightened self-interest by investors and businesses. Responsible investment, in which investors and businesses take account of environmental and social as well as traditional economic factors, has performed strongly in financial terms (the next section) enabling investors to 'do well'.

The financial performance of responsible investment

Financial performance has been documented in the annual benchmarking study conducted by the Responsible Investment Association Australasia,[3] the industry group for responsible investment advisers and fund managers. For the five years ended 2006, the average Australian ethical responsible investment fund returned 14 per cent per year, whereas the average traditional fund achieved 10.9 per cent per year. That is 3 per cent a year better.

The Holy Grail for fund managers is to determine the sources of outperformance – and then to capitalise on them. A factor that causes outperformance is known as 'alpha'. Alpha is the return attributable to a particular variable – in this case corporate responsibility. To calculate alpha, you need to remove the effects of market risk, size of the company, value or growth bias, and momentum. In its search for alpha, AMP in Australia divided the shares of 350 companies into two equally weighted portfolios according to their ranking for corporate responsibility. Over four years, the high-ranking companies outperformed the low-ranking ones by 5 per cent, and over 10 years they outperformed by 3 per cent. This is also reflected in the performance of the AMP Sustainable Future Australian Share Fund, which has outperformed the Australian Stock Exchange (ASX) 200 (the largest 200 companies listed on the ASX) by 3.3 per cent over the past five years.[1]

The bottom line is highlighted by academic research. In 2003, there was an award-winning analysis of 52 studies (involving 33,000 observations) of the relationship between corporate environmental and social performance and corporate financial performance. The analysis concluded that there has been sufficient research to prove a positive association between the two.[4] This demonstrates that responsible businesses do in fact usually make money from their initiatives that make a difference. In turn, responsible investors do indeed do well by doing good.

HOW TO INVEST RESPONSIBLY

Having shown that as an investor you can do well and do good, let us now focus on how to invest responsibly.

Look out for responsible investing

Consumer magazine recommends that you talk to an independent financial adviser before embarking on any new investments.[5] It suggests you go to the Responsible Investment Association (RIA) website[3] for information about RIA certified financial advisers, fund managers and superannuation funds, and look for the certified Sustainable Responsible Investment (SRI) symbol. This symbol is an easy way to tell if a financial adviser, fund manager or superannuation fund is involved in responsible investment and has met the Responsible Investment Association's requirements.

You can use responsible investment strategies to enable you to fund your desired lifestyle and help you fulfil the purpose of making a difference. Traditionally, investment advisers will ask you only about your goals and risk profile in order to determine which investments are most appropriate for you. However, best practice requires that they also ask about whether environmental, social or ethical considerations are important to you. In Australia, the Securities and Investment Commission states that advisers should do this.[6]

Three key questions for a prospective investor who wants to invest responsibly are:

1. What are the important issues for you when considering investment?
2. Are there companies or industries you wish to avoid or invest in?
3. Do you want to foster engagement with companies by investing in them and communicating your environmental and social concerns to companies to encourage their continuous improvement?

The four types of investors

To better understand responsible investment, it is useful to classify investors according to four types: traditional, avoidance, activist and alternative. These are discussed below. In practice, an investor may be a combination of these types, for example, having a portfolio that avoids investing in certain areas, invests in companies and actively encourages them to continuously improve and also allocates money for alternative investments.

A *traditional* investor makes choices of what to invest in based solely on his or her perception of risk and reward and the financial bottom line. Traditional investors do not invest with an eye towards making a difference from an environmental or social perspective. That is not to say, however, that they might not be environmentally or socially responsible in their personal lives, such as contributing some of their time and money to good causes.

An *avoidance* investor wants to avoid particular areas such as tobacco and gambling. They are like the religious groups for whom the term 'sin stocks' was coined to explain how they wanted to avoid investing in areas that were contrary to their religious principles and teachings. Nelson Mandela advocated this approach in the 1970s when he called for investors to avoid investing in companies doing business with the apartheid regime. Mandela saw the resulting divestment as instrumental in overcoming apartheid. This was a powerful example of making a difference.

An *activist* investor wants to engage with companies to encourage them to continuously improve their environmental, social and economic performance. This investor does not believe that responsible investing is limited to the buy and sell decision-making processes, but that it entails dialogue between shareholders and management on a continuing basis. In the USA, considerable emphasis is placed on the benefits to be gained through the shareholder resolution and proxy processes. This is also the focus for a group of the world's largest institutional investors, including the New Zealand Superannuation Fund, who are signatories to the United Nations Principles for Responsible Investment. They have all committed to being active owners, who will exercise their voting rights and engage with companies around environmental, social and corporate governance issues. Another significant local signatory is the ASB Community Trust.

An *alternative* investor wants to invest in alternative initiatives that have a high environmental and social contribution. Alternative investors are willing to sometimes accept somewhat lower financial returns or greater risks in order to generate what they consider higher total returns, by investing in local sustainability initiatives where funding is made available at lower cost to borrowers in areas such as the environment, education, art, healthcare and sustainable agriculture.

Let us test what type of investor you are by using a process from the book *Socially Responsible Investing* (Miller, 1991).[7]

Case 1: The environment matters Imagine you are very concerned about the environmental risk of oil spills and you strongly support the search for alternative energy sources. You own 1000 shares in a small oil company, which is also on the cutting edge of solar energy research. The shares are selling for $5 each.

You have just received a letter from one of the major integrated oil companies that has an especially poor record of oil spills, offering to acquire each of your shares for $15 worth of its own. It is offering to do this primarily to gain control of your company's oil reserves which adjoin its own and which would greatly strengthen its reserves position. It has no plans for the solar energy research unit.

If a majority of the shareholders of your company agree, the acquisition will be consummated and you will receive $15,000 worth of the acquiring company's shares in exchange for your own. If not, you'll keep your shares, which will probably still be worth about $5000.

How would you vote and what would you do if the acquisition were approved?

If you vote in favour of the acquisition and, if it goes through, hold the shares you receive in exchange as an investment in its own right – reflecting the improvement in the acquiring company's reserves position – then you would be classified as a traditional investor.

If you vote in favour of the acquisition, but with the intention of selling the shares you receive for $15,000 as soon as you get them – so that you do not retain an investment in a company with a bad record of serious oil spills – then you are an avoidance investor.

If you vote against the acquisition, and it goes through anyway, and you hold the shares so that you can introduce shareholder resolutions urging that the acquiring company (i) maintain and expand the solar energy research operation and (ii) improve its environmental record by increasing safety precautions in its transportation of oil, then you are an activist investor.

If you vote against the acquisition, and if it goes through anyway, and you sell the shares you receive for $15,000, and look to invest the proceeds in another small company that is researching alternative energy sources, then you are an alternative investor.

Case 2: Social responsibility matters You are very much opposed to smoking but are strongly in favour of corporate support for the arts. You have received a gift of 100 shares in the fictitious company Tobacco International Inc (TII), and apart from its social characteristics, you consider the stock to be an outstanding financial investment. As a leading national corporate supporter of the arts, TII has incorporated a branch of the national art museum into its own corporate headquarters, sponsored major art exhibits nationwide and supported major dance and ballet troupes. It is also a major cigarette manufacturer.

At TII's next annual meeting in two months, you will be given an opportunity to vote on a shareholder resolution urging the company to divest itself of its tobacco business before the year 2015. If that were to occur, the company's cash flow would be reduced substantially and its programmes in support of the arts might have to be curtailed.

If you hold the shares and vote against the resolution in order to indicate your general support of management's business policies, then you are a traditional investor.

If you sell the shares now because you do not want your funds invested in shares of a tobacco company – no matter how good its record of corporate support of the arts – and look for another company that is not in the tobacco business in which to invest your money, then you are an avoidance investor.

If you hold the shares so that you can vote in favour of the company divesting itself of its tobacco business – even if that would jeopardise its programmes in support of the arts – then you are an activist investor.

If you sell the shares and reinvest the proceeds partially in the shares of a small new company that franchises a chain of clinics to help stop smoking, and partially in below-market rate loans to community cultural activities, you are an alternative investor.

Having identified what type of investor you are, responsible investment options that best meet your needs can be recommended according to your preferences towards avoidance, activist and/or alternative investment styles.

Investing locally

Most investors' portfolios include a significant allocation to shares because investing in a diversified portfolio of shares is expected to provide the highest return over time. Many investors use managed investments rather than investing in shares directly as this usually provides higher returns at lower risk. Managed investments offer professional management, more convenience and a better spread of risk.

Let us now look at a New Zealand example, the Asteron Socially Responsible Investment Trust. This was New Zealand's first socially responsible investment trust investing primarily in New Zealand shares and New Zealand's first KiwiSaver Responsible Investment option. In July 2007 when KiwiSaver was launched, it was the only responsible investment option and the only socially responsible investment trust in New Zealand. It is offered by Asteron, part of the Promina Group, which is listed on the Australian and New Zealand stock exchanges. Asteron has partnered with my responsible investment and business research consultancy firm, Rodger Spiller & Associates, to provide guidance and engagement services.

A tool that I have developed for the use of investors and businesses is The Responsibility Scorecard.[8] It provides further insights about the expectations that responsible investors have of businesses in which they invest. The scorecard covers the four 'Ps' of responsible business – Purpose, Principles, Practices and Performance Measurement:

> *Purpose:* How the company defines its purpose, the stated reason for which it exists. Responsible investment encourages companies to explicitly include environmental, social and economic wealth creation.

Principles: The beliefs that guide the company's action. Responsible investors encourage companies to explicitly state and follow a clear set of principles reflecting core values and virtues.

Practices: The actions that a company takes to fulfil its purpose. It is at the level of business practices that the true test of environmental, social and economic wealth creation occurs, as practices reflect the application of purpose and principles.

Performance Measurement: How a company's practices have impacted on its stakeholders, reflected its principles and created environmental, social and economic wealth.

The Responsibility Scorecard includes an inventory of best practices of each of the six main stakeholder groups:

Environmental practices, including reduction, reuse and recycling of materials, energy conservation and environmental audits.

Employees' practices, including effective communication, learning and development opportunities, safe and healthy work environments and equal employment opportunities.

Customers' practices, including industry-leading quality programmes, full product disclosures and safe products.

Suppliers' practices, including long-term purchasing relationships, along with fair and competent handling of dispute.

Community practices, including innovative giving to the community through volunteer programmes, and support for education and job training initiatives, which strengthen the relationship between business and society.

Shareholders' practices, including receiving a good rate of return, comprehensive and clear information and effective management of corporate governance issues.

Some of the responsible business practices in New Zealand can be seen in the following examples of companies that have featured in the Asteron Socially Responsible Investment Trust.

Fisher & Paykel Healthcare's (FPH) Code of Conduct provides a framework of the standards by which the company's directors and employees are expected to live their professional lives. The code is described as an integral part of the company's business, and is reflected in how it tackles business challenges. For example, in response to a large number of job vacancies, FPH established a relationship with Work and Income New Zealand (WINZ), giving its personnel access to the FPH recruitment process and training. WINZ now tailors applications, increasing the chances of job seekers securing work at FPH.

Another practice highly regarded by responsible investors is research and development. Fisher & Paykel Healthcare is a world leader in products that make major contributions in the fields of respiratory humidification, obstructive sleep apnoea, patient warning and neonatal systems, and that are used in hospital intensive care departments. When I did my doctoral research in the mid-1990s, I rated companies on the New Zealand stock exchange, quantifying their strengths and weaknesses in relation to the best practices of responsible businesses, and Fisher & Paykel was the clear winner.

Another example is Fletcher Building, with Corporate Governance Principles that include: 'to promote ethical and responsible decision making'. In relation to the principle 'We value our communities and our environment', they state: 'Our business cares about the world in which it operates. We strive to maintain good relationships with our communities. We act responsibly towards the environment.'[9] A practice reflecting this principle is Fletcher Building's involvement as a partner in the Beacon Pathway Ltd research consortium, which is working to find affordable ways to make homes in New Zealand more resource efficient, cheaper to run, healthier to live in and kinder to our environment. The Waitakere show home, built by the research consortium, uses 30 per cent less energy and 25 per cent less water than similar-sized homes in the area.[10]

Investing internationally

Most investors' portfolios include a significant weighting to international shares to achieve the benefits of increased return and reduced risk that are associated with diversification beyond New Zealand. This is typically achieved through a professionally managed fund.

These funds often invest in responsible investment themes such as clean energy, green transport, water management, waste management, environmental services and sustainable living. These areas are referred to as 'Industries of the Future' based on the idea that these investments are in growth markets largely unaffected by economic cycles, and investors can obtain competitive advantages by spotting trends in environmental technologies at an early stage. Investors can also capitalise on increased consumer awareness about environmental issues, which creates new markets for products such as renewable energy.

One example of clean energy is solar energy. Solar power systems harness solar energy and convert it into hot water or electricity. The global solar energy market has been growing at a rate of 60 per cent per year and is predicted to grow by 30 per cent per year on an ongoing basis.[11] Grid-connected solar technology is now the fastest growing energy technology in the world. Responsible investment funds may invest in solar energy companies.

Another example of a clean energy theme in which responsible investment funds may invest is wind power. This is a rapidly growing form of energy generation in the global electricity generation market; it has averaged 28 per cent annual growth, and is becoming cost competitive with other forms of electricity generation.[11] The expertise for land-based wind farms is well established and offshore wind farms offer significant promise as they benefit from stronger, more consistent winds and less visual impact on communities than land-based ones. Offshore wind farms can use larger turbines, which are more efficient at harvesting wind potential.

An example of a highly regarded, responsible international business is environmental leader Interface, which was included in the 2007 SustainableBusiness.com list of the world's top sustainable shares. Interface is admired for its mission to be a restorative enterprise – putting more into the environment than it takes out. Its founder Ray Anderson refers to this as climbing 'Mt Sustainability' – a challenge he regards as greater than climbing Mt Everest. Its local business, Interface New Zealand Ltd, recently announced that all of its carpet tiles are now climate neutral. Climate neutrality is achieved by measuring the greenhouse gases emitted during the life cycle of the product, reducing them as far as possible and offsetting any remaining emissions. The product life cycle includes raw material acquisition, manufacturing, transportation, use and maintenance and end-of-life disposition. The business is now accounting for all greenhouse gas emissions over the entire life cycle of its carpet. Interface illustrates the growing move of businesses to offer carbon-neutral products and services.

These business initiatives aim to assist with achieving the goal of carbon neutrality. By investing responsibly you too can help to work towards this goal. This step is recommended by Al Gore, in his book and award-winning movie *An Inconvenient Truth*[12] which recommend responsible investing among the actions for individuals to take. Looking after the environment makes good business and investment sense and carbon neutrality is a wise business goal.

A RESPONSIBLE FUTURE

Other examples of responsible business practice can be seen in the Business Ethics Awards created in 1999 by myself, in partnership with Deloitte and the magazine *New Zealand Management*. These are featured on the website of the New Zealand Centre for Business Ethics and Sustainable Development.[13] We created the awards to showcase the kinds of practices that responsible investors look for when they are judging the environmental, social and financial performance of a business. These local examples are used in ethical leadership training such as the course I present through The University of Auckland Business School Executive Short Courses.

I still vividly recall the response from the audience of 700 formally attired business people after the Best Chief Executive, Best Chairman and the other traditional awards had been announced at the 1999 Deloitte/*Management* magazine Top 200 Awards dinner. The MC announced that 'this year we are going to have a new award for business ethics'. There were howls of laughter and many comments about 'business ethics' being an oxymoron. This was the type of response one might have expected from a deep green campaigner. To hear it from the cream of Kiwi business leaders was somewhat sobering. Over the years, the award has become increasingly sought after, reflecting the greater focus on business ethics. No one is laughing now.

A vision for responsible investment in New Zealand in 2020

If I imagine how life could be in New Zealand in 2020 my vision is that by then responsible investing would be the norm. Here is how it would look.

Most investors have their money managed responsibly and enjoy doing well while doing good. The argument that you can have either profit or principle is a distant memory. People look back on this book and marvel that in 2007 there were so many people questioning what in 2020 seems obvious, and wonder why such commonsense was so uncommon back then. Most New Zealanders have a responsible investment plan and portfolio, and their KiwiSaver scheme is in responsible investment.

All investment advisers have training in responsible investment and are able to converse intelligently with their clients on social and environmental matters. They are all certified by the Responsible Investment Association and have the Sustainable and Responsible Investment (SRI) symbol displayed on all their material, testifying to the market that their service systematically addresses environmental, social and ethical considerations. Their newsletters and seminars feature examples of investments that characterise the sustainable and responsible environment, economy and society of 2020.

All fund managers have adopted responsible investment as part of their mainstream investment activity. The New Zealand government followed the lead of the UK and Australia in requiring all superannuation fund managers to incorporate into their investment policy guidelines a position on environmental and social issues, even if a company's position was that it did not consider such issues. And, for all three countries, this was extended to all fund managers. Just as had been the case internationally, this requirement has been remarkably effective in prompting fund managers to recognise their responsibilities in these areas.

The New Zealand Superannuation Fund along with several other crown financial institutions including the Government Superannuation Fund (a retirement fund for government employees) are now, in 2020, world leaders in responsible investment and have helped create a sustainable New Zealand. Having commenced with NZ$2.4 billion in October 2003, the New Zealand Superannuation Fund has now grown to NZ$120 billion.

New Zealand businesses are world leaders in responsible business. The winner of the 2020 New Zealand Business Ethics Award has also received the inaugural Global Responsible Business Award from the United Nations Secretary-General. Most New Zealand businesses have adopted total responsibility management (TRM), enhancing the earlier business focus on total quality management (TQM). Most businesses produce and publish annual Responsibility Scorecards, proudly communicating their purpose of creating environmental, social and economic wealth, their ethical principles and practices. New Zealand businesses welcome the responsible investment revolution as having encouraged them to take responsible business to heart. Business is regarded by most of the community as a powerful force for good.

Responsible investment has had a huge impact at the community level. There is now ample funding for initiatives such as:

- Environmental projects, conservation, recycling and sustainable energy.
- Biodynamic and organic farming, permaculture and other sustainable agriculture initiatives.
- Holistic healthcare, including practitioners and suppliers of complementary healthcare services and products.
- Education encouraging the creative and social development of young New Zealanders.
- Artistic development in many areas.

The Earthsong Eco-Neighbourhood in Ranui, Waitakere City, has flourished from its small beginnings in the 1990s that included loans from responsible investors. It has been emulated throughout the country with the help of responsible investment funds. Another responsible investment success story is the Clean Green Car Company, which used loans from responsible investors to power its business, as have borrowers purchasing hybrid and other environmentally friendly vehicles.

All schools teach responsible business and responsible investment as part of their core curriculum. The Sorted website (http://www.sorted.org.nz) features responsible investing educational material for students. Teachers look back fondly on the trailblazing play *Cent$ational Harry and the Balance of Life* that was initiated in Australia in 2007.[14] In the play, the central character, Harry, discovers that understanding and managing money can be an exciting process, closely tied to emotional and physical wellbeing. This insight was transformational for New Zealand, as students went home and talked to their parents about cash and consciousness.

They have conversations similar to the one I had with my aunt when I was 13. 'Aunty Martha,' I said, 'I need to decide what to study for my options at high school.' 'What do you want to do with your life, kiddo?' she said. 'I want to help make the world a better place,' I replied. 'Well, you had better study money, because money is what makes the world go round.'

The community now recognises the insight from Harvard Business School Professor Moss Kanter, who in 1991 said: 'Money should never be separated from mission. It is an instrument, not an end. Detached from values, it may indeed be the root of all evil. Linked effectively to social purpose, it can be the root of opportunity.'[15]

Back in the 1980s, when I embarked on my responsible investment and business research, the approach was described as pioneering. It took time to gather real momentum, but now in 2020 it has become mainstream, with most investors and businesses doing well and doing good.

ENDNOTES

1 Dennis, A. 2006. *The Imperative for Sustainable and Responsible Investment*. Sydney: AMP Capital Investors.
2 Jones, N. 2006. *The Growing Trend of Consumers Who Care*. Auckland: Nielsen Media Research Panorama.
3 See http://www.responsibleinvestment.org (accessed 28 Sept 2007).
4 Orlitzky, M., Schmidt, F., and Rynes, S. 2003. *Corporate Social and Financial Performance: A meta-analysis*. London: Sage Publications.
5 Bowie, R. 2007. 'Do well while doing good.' Ethical investing report. Consumer, 3 May 2007.
6 To find fund managers adhering to principles of responsible investment, see http://www.responsibleinvestment.org (accessed 28 Sept 2007). The Australian Securities Commission RG 175.110 states: 'Providing entities must form their own view about how far s945A requires inquiries to be made into the client's attitude to environmental, social or ethical considerations. However, as a matter of good practice (and irrespective of any current legal requirement), providing entities should seek to ascertain whether environmental, social or ethical considerations are important to the client and, if they are, conduct reasonable inquiries about them.' See www.asic.gov.au/asic/pdflib.nsf/LookupByFileName/ps175.pdf/$file/ps175.pdf
7 Miller, A. 1991. *Socially Responsible Investing*. New York: Simon & Schuster.
8 See http://www.rodgerspiller.com (accessed 27 Sept 2007).
9 See http://www.fletcherbuilding.com/governance/downloads/Corporate_Governance. PDF (pp. 1 and 126).
10 See http://www.beaconpathway.co.nz/
11 Jupiter Asset Management. 2006. 'The Green Gauge: Clean Energy'. See http://www.jupiteronline.co.uk
12 Gore, A. 2006. *An Inconvenient Truth*. London: Bloomsbury.
13 The website of the New Zealand Centre for Business Ethics and Sustainable Development is http://www.nzcbesd.org.nz (accessed 28 Sept 2007).
14 The play was initiated by the magazine *Ethical Investor*. For more information, see http://www.harry.net.au (accessed 28 Sept 2007).
15 Kanter, R. M. 1991. Money is the root. *Harvard Business Review* (May-June), pp. 9–10.

The sustainable business challenge

Rachel Brown

Rachel Brown is a pioneer in sustainability in New Zealand, with over 15 years experience working with business in sustainable management. She is the founder and Chief Executive Officer of the Sustainable Business Network (SBN), a national network of over 600 members. Rachel is responsible for strategic planning; facilitation; providing advice; brokerage; stakeholder engagement; and partnership development for a range of small and medium-sized businesses, as well as corporate and government agencies. She has a key role in overseeing SBN projects as well as regional development and research into sustainability trends.

Amongst other things, Rachel has run her own small business, has worked in both the private and public sectors and has been on the boards of businesses. She also has experience in the application of The Natural Step Framework as a tool for understanding sustainability. Prior to founding the SBN, Rachel founded and ran Auckland Environmental Business Network.

THE CURRENT SITUATION

The economy of New Zealand, our exporters and key industries are at risk if, as a nation, we do not respond to the call for action around climate change and other key sustainability issues.

The drivers for business to respond to this call are now clear and include: demands from local and international consumers; the importance of our clean green image and signs it is under threat; the negative impact of travel and food miles on businesses that do not reduce their carbon footprint; the power of non-governmental agencies to promote or blacklist companies on the basis of their environmental performance; the long-term savings from investing now in low-carbon, sustainable technologies; and the likelihood of new legislation that will increase costs for companies that continue to use processes associated with high carbon emissions.

THE VISION FOR 2020

New Zealand, with its clean green image, cultural diversity and penchant for innovation, will enjoy a significant piece of the global, low-carbon economic pie. From their unique vantage point, our businesses will be creating innovative products and services within truly sustainable enterprises. These will be greeted with enthusiasm in a global marketplace hungry for sustainability answers. As a nation, we will be leading in renewable energy, eco-tourism, sustainable agriculture and organics, ethical fashion and creativity, eco-tax reform, urban design, sustainable building and much more.

STRATEGIES TO ACHIEVE THE VISION

This is not business as usual! It requires visionary thinking and innovative solutions. In developing a business, owners need to consider how the business will contribute to a 'sustainable world'. It is vital to be aware of the current constraints and challenges facing us, such as declining water resources, air quality and biodiversity, and increasing pollution, population demands and inequality.

The Natural Step Framework is a useful tool for developing a business vision and for providing priorities for action. This can be complimented with the Get Sustainable Challenge, which encourages business to think about how sustainability can be achieved through leadership and commitment, future thinking, careful resource use, appropriate systems and processes, product and service design, relationships and communications. Businesses need to take advantage of the growing number of other tools available to verify claims or meet industry best practice standards like the well-developed carboNZero programme. There is a range of good community-based, but non-certified, carbon offsets too.

> Sustainability is now the business imperative of the century. Developing strategies to take on sustainability is vital for business and New Zealand's long-term survival.

THE TIPPING POINT FOR BUSINESS

Ecological and social issues are becoming more important than ever and consumers are calling for action. We have reached the sustainability tipping point and sustainability is now the imperative for business this century.

It is only recently that more businesses have started to sit up, listen and seriously rethink their role in solving the big ecological issues of our time, such as climate change. The following is a discussion of key drivers that businesses must take into account if they want to be leaders in the sustainable economics of the future.

Sixty per cent of ecological systems are being degraded

Major problems such as pollution, deforestation, species loss and global warming are all side effects of the activities that provide us with food, transport, shelter, clothing and the seemingly endless array of consumer goods on the market today. It is now clear that the planet's ecosystems cannot sustain this style of consumption for much longer.

Key members of the scientific community have been trying to draw the attention of business to these issues for decades now. Until recently, this call has been heard by only a very small number of passionate CEOs committed to playing a positive role in ecological restoration through their businesses. For the vast majority of business leaders, these problems have been left for others to solve, perhaps because they have failed to understand what is at stake. Maybe, too, these leaders have lacked the imagination to see that profitable business and ecological ethics are in fact highly compatible.

The release of the *2005 Millennium Ecosystems Assessment Report*[1] goes someway to redressing business's lack of understanding in what is at stake. The report provided that much needed scientific consensus and a better understanding of the state of the planet. It warns that approximately 60 per cent of what is needed to support life on Earth (such as fresh water, clean air and a relatively stable climate) is being degraded or used unsustainably. In the report, scientists warn that harmful consequences of this degradation to human health are already being felt and could grow significantly worse over the next 50 years.

Our changing climate is already playing havoc on business performance, particularly businesses in sectors such as agriculture, transport and insurance, and has the potential to impact on tourism. Concern is increasing, particularly in the UK and Europe, about air travel and its contribution to climate change.

An Inconvenient Truth and rising public concern

Humanity is sitting on a ticking time bomb. If the vast majority of the world's scientists are right, we have just 10 years to avert a major catastrophe that could send our entire planet into a tail-spin of epic destruction involving extreme weather, floods, droughts, epidemics and killer heat waves beyond anything we have ever experienced.

So starts the global warming film of the century – *An Inconvenient Truth*.[2] Al Gore's film enlisted the world's people in his crusade, informing them about the need to halt global warming in its tracks. He did it beautifully by exposing the myths and misconceptions that surround global warming, which he calls the biggest moral challenge facing our global civilisation.

Thousands of Kiwis have seen *An Inconvenient Truth* and been motivated to take action. Research on New Zealanders' attitudes to economic growth, published well before (April 2004)[3] the movie's release, demonstrated that even before this new increase of awareness, Kiwis have tended to share some basic common values. We care greatly about our quality of life, the environment, health and education, about fairness, race relations and innovation. We want to see these values incorporated in the way we do business, and fear the social and environmental impact of economic growth generated by 'business as usual'.

In 2005 and again in 2007, Moxie Design Group researched the New Zealand 'solution seekers' market. Moxie describe solution seekers as Kiwis who are actively interested in seeking out products such as environmentally friendly building supplies, socially responsible investing, alternative healthcare, organic clothing and food, personal development media, yoga and other fitness products, and eco-tourism.[4]

In 2007, solution seekers made up 32 per cent of the New Zealand market. In the USA, they are described as a US$209 billion market (see Figure 1). This group of Kiwis has grown by 6 per cent in the last two years and will have grown again since the release of *An Inconvenient Truth*.

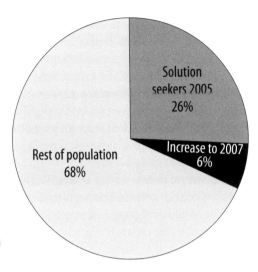

Fig. 1 2005–2007 Solution Seekers

Of course it is not just consumers that care about the environmental and social ethics of a business – staff do too. By taking on 'values'-based issues, staff feel more compelled to support the business and demonstrate a higher level of commitment and productivity. And for most people working inside businesses, ongoing motivation is not all about money. In terms of recruitment, given the growing interest in environmental and social issues, demonstrating alignment with those issues also helps attract talented and committed staff, especially young people and those seeking to work for the greater good.

Clean green New Zealand: under threat?

Over the years, New Zealand businesses have benefitted from our clean green image. In fact, New Zealand depends on this image for the majority of its export base and therefore its economic and social wellbeing. This has particularly been the case for our food and beverage sector, and the image is a powerful pull for tourists wanting to experience '100% Pure New Zealand' (the name of a Tourism New Zealand campaign).

Unfortunately, it appears that clean green New Zealand is a result of good luck rather than good management – and that luck may run out, taking our natural advantage with it. The OECD's 2007 *Environmental Performance Review of New Zealand*[5] highlighted some embarrassing figures on our poor performance in some fundamentally important environmental areas – particularly in the protection of waterways, waste management, energy efficiency performance and climate protection efforts. This poor performance leaves us at risk of squandering this global image with potentially catastrophic consequences, especially for the primary production sector.

Sustainability is vital not only to our national identity, and the wellbeing of our families and communities, but also to securing the future prosperity of our country through thriving businesses.

Food and travel miles: a new trade barrier?

As countries and corporations grapple with how they are going to respond to climate change, concepts such as 'food miles' have surfaced. Food or travel miles measure the distance that food, products or people travel and the energy consumption of the transportation mode involved (for example, ships use less energy per unit of weight moved per kilometre than planes). Crude measurements that do not take account of transport mode assume that 'the greater the distance, the greater the environmental impact'. As a concept, food and travel miles have gained traction with a number of environmental organisations, the media, as well as some of New Zealand's key business customers.

As a result of our geographic isolation, travel miles now affect all exporters and may become a major trade barrier. They are an easy concept to advertise and for the public to grasp, and are particularly an issue for markets in the USA, UK and Europe, though less so for Australia (but this too is changing).

We know that international buyers are willing to pay a premium for products and services that are produced in a sustainable manner and/or are carbon neutral. Businesses that can demonstrate this will be able to stay afloat and be positioned at the more lucrative end of overseas markets. So, if businesses are to continue to trade on New Zealand's clean green image, we will have to ensure that our clean green image continues to stand up to international scrutiny and that we can truthfully demonstrate that buying New Zealand-made will contribute to a sustainable future for all.

Active NGOs

People power is on the move. Non-governmental organisations (NGOs) have begun to measure the environmental and social performance of businesses in a very public way. For example, over the past few years,

Greenpeace Aotearoa New Zealand has produced a number of consumer awareness campaigns on issues such as genetic engineering (with the *GE Free Food Guide*[6]) and renewable energy.

In the case of renewable energy, Greenpeace Aotearoa New Zealand developed the *Clean Energy Guide*[7] where they attempted to rank electricity companies by their commitment towards providing renewable energy. Companies were put into three categories on the basis of how they generate electricity, their generation plans and their policy on future energy development. There is a visual thermometer that ranks all companies accordingly with the least favoured in red at the top through orange to green companies at the bottom.

This created outrage in those providers in the 'red' zone and increased market share for those in the 'green'. This action from the NGO sector highlights one of the biggest fears of any business – no one wants to be shown to be a bad corporate citizen.

A Stern look at the economic imperative

Of course, there are cost implications in a shift towards a more sustainable economy. Thankfully, economic evidence suggests the cost of not changing is likely to be significantly higher.

The recently released *Stern Review of the Economics of Climate Change*[8] estimated that unabated climate change could cost the world economy between 5 per cent and 20 per cent of GDP each year. At the same time the cost of reducing emissions for the same period could be limited to just 1 per cent of global GDP if governments are willing to use financial mechanisms to encourage businesses and individuals to seek out alternatives that are less carbon-intensive.

According the to *Stern Review*, each tonne of CO_2 emitted causes environmental damage worth at least $US85, but emissions can be cut at a cost of less than US$25 a tonne. Shifting the world onto a low-carbon path could eventually benefit the global economy by US$2.5 trillion a year.

By 2050, markets for low-carbon, sustainable technologies could be worth at least NZ$500 billion. New Zealand, with its clean green image, culturally diverse and creative population and penchant for innovation, could enjoy a significant piece of the low-carbon economic pie. Our image provides businesses with a unique vantage point from which to create truly sustainable enterprises with innovative product and service solutions for a global marketplace hungry for sustainability answers. As a nation we could be leading in renewable energy, eco-tourism, sustainable agriculture and organics, ethical fashion and creativity, eco-tax reform, urban design, green building and much more.

A boost in political support and incentives

Following on from a year of increased public awareness has been a growing recognition of the importance of the path to a more sustainable society. The Labour government committed the country to be 'the first nation to be truly sustainable, and aim to be carbon neutral' (Rt. Hon Helen Clark).

The 2007 budget has seen Labour commit to funding a number of sustainability-oriented policies. Around NZ$800 million is committed to a programme of action that covers a range of infrastructural areas, research on energy-efficient technologies, sustainable government procurement, enhanced eco-verification, and shifting the public service towards carbon neutrality. For those businesses that are new to sustainability, there is now a plethora of government-funded programmes, and grants, aimed at increasing the capability of firms to become more sustainable. And incentives are now being delivered through government. Of particular interest to business is the changing purchasing requirement for more environmentally friendly product attributes or systems from suppliers.

'Govt3' is the programme that helps central government agencies become more sustainable.[9] A key part of the programme recognises that strategic government spending can lift the environmental performance of businesses. New Zealand's government spends at least $25 billion per year on goods and services; it is responsible for more than 30 per cent of the buildings in New Zealand and is a large purchaser of information technology, paper and vehicles. Government, along with a growing number of large corporations, such as IAG insurance, ACC and others, have now adopted sustainable procurement requirements for their suppliers. This is a strong signal to suppliers that sustainability will shortly be an entry level requirement to do business with the big purchasers in the future.

This shift also signals the potential for more environmentally friendly legislation, such as the Waste Minimisation (Solids) Bill currently under consideration, which proposes actions to support waste reduction including provision for waste levies, setting targets for reducing waste in landfills and cleanfills, and providing for product stewardship programmes. Other options currently being explored include the introduction of an emissions trading scheme, a greenhouse gas charge and emission reduction agreements, along with a host of recommendations for improving the efficiency of the country's vehicle fleet.

In addition, government is increasing its support for pioneering organisations that have been pushing for change in business practices such as the New Zealand Business Council for Sustainable Development, The Sustainable Business Network and the New Zealand Green Building Council. A focus on key sectors (for example, tourism, aquaculture, building) will also see the development of new industry-lead sustainability-focused accreditation systems and standards.

GOING CARBON NEUTRAL AND THE SUSTAINABILITY CHALLENGE

The business opportunities that arise by aiming for carbon neutrality are clear: enhanced reputation, long-term cost savings and increased productivity, staff loyalty and staying ahead of regulations. It is also easier to avoid being seen as old-fashioned and out of touch with the future, than having to suddenly change when new regulations come in force to protect the environment, and being boycotted by solution-seeking consumers and ethical investors.

There are three main steps to going carbon neutral:

- Determine your carbon emissions by using a carbon calculator
- Reduce your emissions as much as you can
- Offset what is remaining.

It is important to note, however, that although going carbon neutral is a good start, the Sustainable Business Network recommends a much broader sustainability strategy. So, beyond looking at conventional issues around energy, waste, water, transport, office supplies etc, businesses integrating sustainability strategies need to embark on a much more encompassing, transformative journey; one in which they invest in cleaner products, undertake product life-cycle analyses and implement mega efficiencies in resource use. If we are to reach our goal of a truly sustainable, carbon neutral-nation we require such a shift within business.

By understanding the ecological and social issues we are facing, by thinking of solutions 'outside of the norm', we open ourselves up to developing the leading edge innovative sustainable solutions for a global market. One step towards this is developing a much greater understanding of the impacts your business is having on the world. Ideally, those running businesses would be spending more time learning about the global ecosphere so they have sufficient knowledge and understanding to effect change for the better.

Many involved in business, particularly economists, believe that the global economy is the system of which everything else is a part, everything else being people, the air, land and water and all living things on the planet. The economy from this perspective apparently can run on a continuous growth path. Unfortunately, this is simply not the case.

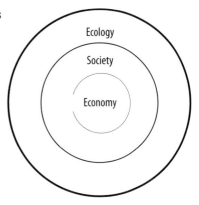

Fig. 2 Sustainability model

As Figure 2 shows, the economy is simply a subset of society – it is a system designed by people for the exchange of goods and services; and society is a subset of the broader ecological system or ecosphere, which includes all the things we rely on for our survival.

There are a growing number of whole-system approaches being used now to meet the challenge of sustainability. The two that are better known are Natural Capitalism and The Natural Step (TNS) Framework, which both provide a set of simple principles needed to live within our ecological limits: using resources more efficiently and fairly; redesigning all products and services using innovation inspired by nature; managing all systems to be restorative – not just to the environment but to society too.

THE NATURAL STEP FRAMEWORK

In 1989, Professor Dr Karl-Henrik Robèrt,[10] developed a framework for society to better live within its ecological limits. TNS Framework's development was based on a consensus among leading scientists and key figures in Sweden, and TNS Framework sets out the system conditions for sustainability.[11] The framework has been applied to fields from small design companies to green building to the management of cities and has established offices across the globe, including here in New Zealand. Robèrt's system conditions are significant in that

> **Phoenix Organics: pioneers with TNS Framework in New Zealand**
> Phoenix Organics (previously Phoenix Natural Foods) was the first company in New Zealand to adopt TNS Framework and it has made many changes towards being environmentally and socially sustainable, while becoming an increasingly profitable company.
> The benefits of using TNS Framework for Phoenix, their environment and their community, have included increased sales, reduced costs through reduced waste and sourcing locally-made products, greater customer loyalty, increased support for organic food production, reduced impacts on local ecosystems, restoration of native biodiversity (while offsetting fossil carbon emissions), enhanced availability of clean and healthy food, more unified staff and additional employment, and great personal satisfaction for the company owners and staff.[12]

they are scientifically-based, but readily understandable, principles for sustainability. The framework consists of four 'system' conditions for life, three about the human interaction with the environment and the fourth about society.

In a sustainable society, nature is not subject to systematically increasing:

- Concentrations of substances extracted from the Earth's crust. Since the industrial revolution we have been extracting raw materials, such as metals and fossil fuels, from the Earth's crust at a rate much greater than the rate that they (including fossil fuel-derived carbon emissions) are able to be assimilated safely back into the environment.
- Concentrations of substances produced by society. All over the world, we are creating larger amounts of chemical compounds, which get into the environment. In a growing number of cases these are impacting on the overall health of people and biological systems that cannot render them inert (for example, PCB's or nitrogen oxides).
- Degradation by physical means. We are now feeling the effects of serious degradation of the stocks of key resources such as fisheries and native forests, and of poor management of others such as wetlands and our soils.

And the people of a sustainable society are not subject to:

- Conditions that systematically undermine their capacity to meet their needs. Ecological and social systems are tightly linked, and social pressures will continue to rise if we continue to remove some people's right to the basics of life (such as clean air, water, food, shelter). With growing populations, and increasing competition for resources, we are likely to impact more on natural systems in order to meet our immediate basic needs. Unless these issues are addressed we will see a growth in conflict, poverty and other forms of social stress.

These four 'system' conditions are visually represented in Figure 3 (below). The arrows show the positive and negative flows associated with the movement of resources between the Earth's crust, the ecosphere and society.

RELATING TNS FRAMEWORK'S 'SYSTEM' TO BUSINESS

For business leaders, the real task is now in translating this information about the 'system' and its 'conditions' into strategies that will future-proof your business. This next section attempts to explain the steps that a number of businesses around the world have taken to integrate sustainability through TNS principles into the fabric of their business operations.

TNS has a number of tools that are useful for an understanding of how well your business is doing (or not doing) in terms of the conditions needed for a sustainable society. The 'funnel' is a useful tool, and is available on the TNS website,[11] for assessing your current performance and any future or potential risks for your business. The resulting information allows you to create a list of all of your business's current practices that contravene the four conditions.

In this next section I will refer to a fictitious courier company, 'Sussed Couriers', as an example of what is possible for a mainstream business. The sustainability issues and drivers that the owners of this company would need to think through strategically might include a decline in cheap petrol, declining air quality, decreasing climate stability (increased storm events), increasingly strict emission and vehicle standards, increasing compliance costs, greater pressure from clients to offset carbon emissions, a demand for more equitable pay for staff and contractors, increased innovation from competitors and higher insurance premiums.

THE SUSTAINABLE BUSINESS CHALLENGE

Fig. 3 The four system conditions

Developing a vision of success

A carbon-neutral business needs a vision. This vision must fit within the systems conditions provided by TNS Framework. The aim of the exercise is not to be limited by current constraints or to think about ways to recreate the business so that it could address various drivers of change. This is a time for creative solutions. The example below presents the vision of Sussed Couriers.

'Sussed Couriers'
The world's first restorative, carbon-neutral courier company.
For Sussed Couriers this means:

- Running the fastest cycle fleet in town;
- Operating the cleanest, greenest most efficient vehicles out of town;
- Living in the greenest, most efficient building and operating the most sustainable courier business;
- Offsetting any net carbon emissions and restoring nature's diversity through tree planting.

All while meeting – or even exceeding – the needs of customers and staff.

Once the vision is developed, the next step is to agree on the high-level priorities that will help move the business toward the vision. From years of working with businesses, the Sustainable Business Network (SBN) has developed a set of eight characteristics, the use of which makes planning and delivering the vision easier. These characteristics are important to achieving and maintaining long-term commitment to sustainability.

- Top-level commitment is crucial for driving action towards the vision
- Company culture and decision-making need to support the call to action and motivate and empower staff to act
- Future thinking ensures that the system conditions for a sustainable future continue to be assessed and reviewed
- Choice and use of resources is critical for ensuring the most effective use of resources
- Business systems and processes must support the most efficient and effective use of key resources: staff, raw materials and prevent wastage
- Considered design and creation of product and services is essential to meeting the needs of markets locally and globally for sustainable products and services
- Relationships with organisations across sectors support the work being done and businesses are able to use their power to influence others – particularly suppliers
- Communication of what the company is doing and seeking feedback on progress is useful.

Cycle of continuous improvement
Improvement cycles help businesses easily progress in simple, manageable steps (see Figure 4). We recommend a simple four stage improvement cycle starting with the first stage of formalising; stage two involves measuring and monitoring, stage three prioritising and acting, and the fourth stage is reporting on progress. There is no set timescale for moving from one stage to the next, although it is worth aiming to conclude an entire cycle within 12 months. Once the first 'improvement cycle' has been completed, many of the necessary systems will be in place. Of course, improvement will always rely on staff commitment at every level, but assuming being a carbon-neutral business is

an ongoing priority, with every 'improvement cycle' the sustainability performance will increase.

It is really important that the initiation and progression of this cycle be overseen by a dedicated individual or team. For larger companies, this team should be multi-disciplinary and drawn from both the strategic and operations levels. In addition to this team, it is recommended that as many staff as possible are involved at every stage. And gaining senior commitment to this approach is critical to ensure that it is part of the company's strategic direction.

1 FORMALISE →	2 MEASURE AND MONITOR ↓
• **Year 1:** Formalise your approach in a policy or document and ensure there is ownership of this • **Onwards:** re-establish your commitment, review policy, communicate to staff • **Involve:** A champion from management	• **Year 1:** Establish baseline, automated data collection and monitoring systems • **Onwards:** Measure progress against baseline using monitoring systems • **Involve:** Wider staff
4 REPORT AND REWARD ←	3 PRIORITISE AND ACT ←
• **After 6–12 months:** Communicate and celebrate improvements against targets, eg. Annual Report, staff meetings • Make relevant links to external reporting/certification progammes • **Onwards:** Include in performance reviews • Involve: HR/reporting/communications	• Identify areas for improvement (minimise impacts > maximise opportunities) • Prioritise • Set new targets for each priority • Action plan by determining steps towards achieving each target • Involve: CEO/GM/top-tier managers and wider staff

Fig. 4 Cycle of continuous improvement

Formalise In the first year, it is important that you formalise your business's approach to carbon neutrality and sustainability (be that in transport, energy use, waste, community involvement) in the form of a policy or written document. This will signal that carbon neutrality and sustainability are considered essential parts of your business, enabling the approach to enjoy a higher level of accountability (through, for example, measurable outcomes) and preventing the approach from being abandoned with staff changes.

The policy or written document should describe the values and vision of the business's approach to carbon neutrality and sustainability. For example, as Sussed Couriers is focusing on efforts to reduce transportation impacts, the policy may outline specific commitments, such as offering a travel allowance for bike travel, offering loans for staff to purchase bikes, providing staff with tickets for public transport options, ensuring that any cars purchased and used by the company are the most efficient for the role that is needed.

Time should be taken during staff meetings to introduce the approach, and any subsequent changes to it, to encourage company-wide ownership of the programme. Displaying any acted-on commitments publicly, so that they become more than just words on a written document, is also useful.

In subsequent years, it is useful to refresh and re-establish the business's commitment to its approach, then to review the policy and make modifications to reflect any changes or increasing commitments.

Measure and monitor Once the policy is formalised, the next stage is to get baseline information about your business from which you will be able to measure your business's performance and chart future progress.

For Sussed Couriers, assessing or measuring their current situation would first involve determining their carbon footprint. This would require them to gather information on all carbon related activities such as travel, energy and waste; and then input this information into a carbon calculator thus providing a baseline from which is measure and report on progress.

The main method for measuring carbon emissions involves using a 'carbon calculator'. See the Carbon calculators section of 'Tools to support the delivery' at the end of this chapter.

There are other tools available that can be used to develop an ongoing monitoring system to measure the impact of your actions (for example, the Catalyst carbon calculator provides a free and useful series of spreadsheets for monitoring the key carbon-related emissions). You will need to report the results at staff meetings to encourage further action and improvement. Keeping track of this information will help you to identify whether you are on track to achieve your target and also provide you with a general picture of how successful the initiative is within your workplace.

Prioritise and act Each action the business undertakes should be linked to its stated policy or vision. Use your policy document from Stage 1 and your baseline information from Stage 2 to identify your priority areas for implementation. The following table lists actions relevant for our fictitious company Sussed Couriers.

	SUSSED COURIERS' ACTION LIST – LISTED IN PRIORITY ORDER	
1	Establish baseline	Use the Catalyst Carbon calculator to establish our baseline data and recalculate progress using the calculator every 6 months.
2	Review vehicle fleet	Review all vehicles in the fleet and ensure all are the most efficient option. Introduce system for upgrading our vehicles.
3	Encourage better staff travel options	Introduce incentives to motivate staff to use public transport, or to cycle or walk.
4	Increase cycle fleet by 5%	Assess opportunity to increase number of cycles in fleet given increased traffic.
5	Incentives for customers	Work with customers to identify incentives to encourage the growth in cycle fleet.
6	Use latest technology	Use technology to maximise efficiency of fleet.
7	Community programmes	Work with local cycle action network to assist in developing safe cycling in the area.
8	Offset carbon emissions	Join the SBN's GreenFleet to remain informed on issues, and offset carbon emissions by involving staff in their regular tree planting programmes.
9	Communicate what we are doing	Develop an on-line reporting system and ensure customers, staff and stakeholders are aware of what we are doing by including information in our newsletter.
10	Enter awards	Participate in the Get Sustainable Challenge and enter at least one award competition.

Once the list is developed you must prioritise the actions and allocate responsibility and timeframes. After your prioritisation process, set a target for each priority, which the business as a whole can aim for during the next 12 months. Record and publicise this target and its associated key performance indicator.

Define the steps necessary for you to achieve each target. This exercise is best done as a group that includes staff from all levels of the business. Start by identifying possible barriers to the business achieving a particular target, and then develop ways to overcome them. Determine the steps you will need to take so that the business can reach its target within 12 months.

Report and reward Reporting does not have to be an onerous task! Planning is the key to ensuring that it is as simple as possible. At least one person should be overseeing the whole improvement cycle process, and a coordinated approach can be used to track progress against each target during the year. Different people can oversee individual targets if necessary.

Report results either every month or quarter to provide feedback and motivation to staff, including management. This will also help staff to develop a better understanding of the scope of what the business is attempting to do. In addition, it will encourage participation as it shows that the business is serious about becoming sustainable through the number of sustainability issues it is addressing. At the end of a 12-month period, you should have a set of final results that will allow you to determine whether your business has achieved its targets.

You will need to decide upon an external and internal means of communicating and reporting your annual results, such as in an annual report or a monthly newsletter. Communicate success stories via email or through pictures on central notice boards to encourage staff to continue with their progress.

Celebrate your achievements. You could celebrate collectively by going out to lunch or you could build rewards into your performance appraisal process. Remember, keep an eye on the bigger picture; targets are just a tool to help in your overall improvement process. To keep everyone motivated, consider entering an awards programme such as the SBN's Sustainable Business awards.

TOOLS TO SUPPORT THE DELIVERY
There is a growing range of environmental management systems, industry standards and certification systems on offer in New Zealand. All are useful tools for supporting the implementation of sustainability. They are all compatible with TNS Framework outlined in this chapter, but are relevant to different industries. The key ones are listed below.

Get Sustainable Challenge
The Get Sustainable Challenge is for business, communities and organisations:

- That want to be more sustainable, but need some help beginning the journey
- That are already changing their business practices and want to take the next step
- That are well advanced in sustainability, but want a 'tune up'
- And anyone who wants advice on how to be more sustainable, future-proof their business and add to their bottom line.

The Get Sustainable Challenge was developed by the SBN.[13] We developed the Challenge because we felt that there were no support programmes in New Zealand that could help businesses understand sustainability and reframe themselves in line with it. The Challenge begins with a personal assessment of your business by SBN staff. This looks at the sustainability of your whole business, from strategy and policies, to operations and performance monitoring. We will work with you to identify your specific challenges in improving your business's sustainability and to help you to keep moving forward throughout the year. We have developed dedicated resources and workshops to provide each participant with a tailored support.

Each participating business has the opportunity to enter the annual Get Sustainable Challenge awards where businesses are recognised regionally and nationally for their efforts in sustainability.

Carbon- and transport-related management systems
The carboNZero programme, referred to earlier, can be used not only to measure your business's emissions but also to offset any remaining, unavoidable emissions. The programme has been created by Landcare

Research Manaaki Whenua especially for New Zealand. Organisations can seek carboNZero certification for their efforts to reduce their greenhouse gas emissions and to offset (balanced or neutralise) their remaining unavoidable emissions.[14] They do the latter by working with Landcare, which under the carboNZero scheme has a formal relationship with the QE2 Trust, to fence off private land, under covenant, to allow for natural regeneration and to monitor and report on progress of the regeneration.

As part of the SBN's ongoing commitment to sustainability, and to demonstrate that taking responsibility for one's greenhouse gas emissions is straightforward and inexpensive, we developed GreenFleet – a programme for reducing and offsetting carbon emissions for company fleets, individuals and flights.[15] GreenFleet is focused on reducing emissions, opting for cleaner burning fuels and vehicles to improve efficiency and reduce pollutants. Offsetting what is left is tackled by a Memorandum of Understanding that GreenFleet has with a small number of established community-based planting partners who undertake to plant a certain number of trees on behalf of GreenFleet members. Twenty-three trees are planted for each vehicle. Although GreenFleet offsets are not certified, planting partners must provide reports on the number of stems planted, the species planted and commit to manage the tree maintenance for at least two years.

Carbon calculators There are a large number of these available globally and a few options suitable for New Zealand. Catalyst R&D has developed one that SBN uses.[16] Or you can use the household emissions calculator developed by Landcare Research Manaaki Whenua as part of its carboNZero programme.[17]

Environmental management systems
These are systems adopted by businesses that promote continual improvement in environmental performance. They typically involve environmental policy development, performance monitoring, audits and reviews to ensure that policy aims and objectives are being met.

ISO 14001 is part of the International Standard Organisations set of standards. It is the one that focuses on Environmental Management Systems and is the preferred standard for many large businesses around the world. The aim of ISO 14001 is to protect and prevent pollution in the environment while balancing that against social and economic realities. It requires that businesses have an environmental policy and that they have procedures in place to establish and assess the effectiveness of their activities to reduce their environmental impact.[18]

Enviro-Mark® is a health, safety and environment management package for businesses developed by Landcare Research Manaaki Whenua.[19] It is aimed at:

- Reducing energy consumption
- Reducing waste to landfill
- Reducing potential legal compliance risks
- Raising staff awareness
- Achieving health, safety, and environmental legal compliance
- Reducing environmental risk
- Increasing resource use efficiency (i.e. water, waste, energy, and raw materials)
- Improving status with stakeholders by demonstrating environmental commitment.

Enviro-Mark® is also useful for purchasing staff seeking to buy products from environmentally responsible producers or suppliers.

Environmental/ Industry Standards and labels

There is a growing number of environmental or industry standards around, some of which are listed below. There are also product labels that are sometimes developed by industry bodies. A product label identifies a preferred product or service in a category, on environmental grounds, and is based on life cycle considerations. In contrast to 'green' symbols or claim statements developed by manufacturers and service providers, an eco-label is awarded by an impartial third party to certain products or services that have been determined to meet environmental leadership criteria.

Environmental Choice is New Zealand's eco-labelling programme that, like ISO 14001, sets standards for products to met.[20] It is based on an independent assessment of the environmental advantages of products and is endorsed by the Ministry for the Environment.

There is a range of organic certification systems in New Zealand. The main ones are listed overleaf:

Biogro http://www.bio-gro.co.nz
AgriQuality http://www.agriquality.co.nz
Demeter http://www.biodynamic.org.nz
OrganicFarmNZ http://www.organicfarm.org.nz
IFOAM. www.ifoam.org

Industry standards and certification schemes include set standards that require specific environmental criteria to be met. Schemes for five important sectors are:

Tourism: GreenGlobe21 www.greenglobenz.com/
Forestry: Forestry Stewardship Council (FSC) www.fscus.org/
Marine: Marine Stewardship Council (MSC) www.msc.org/
Wine: Sustainable Wine Growing www.nzwine.com/swnz/index.html
Building: GreenStar rating www.nzgbc.org.nz/

CONCLUSION

The drivers for business are clear – the change in thinking required is radical and the associated challenge for business is significant. It is going to take a whole new approach to make being sustainable work. New Zealand's overall wellbeing and prosperity depend on it. If we do not respond quickly, I believe that as a nation we will fail.

If we have the will, we can make it happen – and quickly. We are ambitious people with a natural mix of determination and creative thinking. We can come up with solutions that draw on our diverse cultures and common values, which are strongly linked to sustainability and ecological principles. We have the opportunity to base our economic future on principles such as quality, integrity, effectiveness and natural beauty. Some businesses are already doing this and doing it very well. Their experiences could be used to inspire and their key staff could mentor new business leaders ready to take on these issues.

We are now part of a global culture where one country and one product looks ever more like another. As Rod Oram put it, 'we could become the first country to earn a First World, sustainable living from our natural environment; through myriad small, entrepreneurial and international companies; supporting and being supported by a just, multicultural society.'[21]

ENDNOTES

1. This was the first of several reports commissioned by the United Nations Environmental Protection Agency. The assessments were carried out 2001–2005 and involved 1,360 experts worldwide. For more information, visit http://www.millenniumassessment.org/ (accessed 30 Sept 2007).
2. *An Inconvenient Truth* (motion picture). 2006. Paramount Classics and Participant Productions.
3. The research was carried out by the government's Growth and Innovation Advisory Board. The full report is available at http://www.giab.govt.nz/work-programme/growth/research-summary/index.html (accessed 30 Sept 2007).
4. *Understanding the Market for Sustainable Living: The growth in the New Zealand Solutions Seekers Market.* February 2006 Moxie Design Group Limited. p3. http://www.moxie.co.nz (accessed 30 Sept 2007).
5. This document can be downloaded from http://www.oecd.org/document/10/0,3343,en_2649_34307_37915274_1_1_1_1,00.html (accessed 30 Sept 2007).
6. Greenpeace Aotearoa New Zealand. 2007. *GE Free Food Guide*. You can read more about the guide, and download a copy, at http://www.greenpeace.org.nz/truefood/default.asp (accessed 30 Sept 2007).
7. Greenpeace Aotearoa New Zealand. 2007. *Clean Energy Guide*. You can read more about the guide, and check out your energy provider's performance, at http://www.cleanenergyguide.org.nz/ (accessed 30 Sept 2007).
8. Stern, N. 2007. *Stern Review of the Economics of Climate Change*. Cambridge: Cambridge University Press. http://www.hm-treasury.gov.uk/Independent_Reviews/stern_review_economics_climate_change/sternreview_index.cfm (accessed 11 Sept 2007).
9. Information about this Ministry for the Environment programme is available at http://www.mfe.govt.nz/issues/sustainable-industry/govt3/ (accessed 30 Sept 2007).
10. Karl-Henrick Robert MD PhD is one of Sweden's leading cancer scientists and a key figure in the worldwide sustainability movement.
11. To find out more about The Natural Step Framework, visit the website which has excellent resources and links: http://www.naturalstep.org.nz (accessed 30 Sept 2007).
12. Phoenix Organics website: http://www.phoenixorganics.co.nz (accessed 30 Sept 2007). See also: http://www.naturalstep.org.nz/a-t-f-case-studies.asp#phoenix
13. Find out more about the Get Sustainable Challenge by visiting http://www.sustainable.org.nz/getsustchallenge (accessed 30 Sept 2007).
14. To learn more about the carboNZero programme, visit http://www.carbonzero.co.nz (accessed 30 Sept 2007).
15. To find out more about GreenFleet, visit http://www.sustainable.org.nz/greenfleet (accessed 30 Sept 2007).
16. Go to the Catalyst website, to the Catalyst Carbon Services page and then the Annual Carbon Emissions (ACE) page to download its carbon calculator. http://www.catalystnz.co.nz (accessed 30 Sept 2007).
17. To access the household carbon emissions calculator, visit http://www.carbonzero.co.nz (accessed 30 Sept 2007).
18. To order a copy of ISO 14001, go to the Standards New Zealand website, http://www.standards.co.nz (accessed 30 Sept 2007).
19. More information about Enviro-Mark (R) is available at http://www.enviro-mark.co.nz (accessed 30 Sept 2007).
20. Find out more at http://www.enviro-choice.org.nz (accessed 30 Sept 2007).
21. See: http://www.anewnz.org.nz/vision.asp?id=16 (accessed 30 Sept 2007).

Carbon neutrality and the law

Klaus Bosselmann

Professor Klaus Bosselmann, PhD, is Director of the New Zealand Centre for Environmental Law at the University of Auckland. He has been a consultant on climate change law to the European Commission, the German Government and non-governmental organisations. His publications on New Zealand's climate change policies include *Climate Change in New Zealand* (2002, with Jim Salinger and Jenny Fuller) and a number of journal articles and reports.

THE CURRENT SITUATION
New Zealand has the world's most advanced environmental law, the Resource Management Act, but it is not used for reducing our greenhouse gas emissions. In 2004, the government deliberately exempted greenhouse gases from the Act. Therefore, no legislation is in place, at present, to reduce carbon emissions. The recently announced emission-trading system and other proposed policies are insufficient to meet our Kyoto target, let alone to achieve carbon neutrality. The sad reality is that, per person, New Zealand is among the highest emitting countries in the world and our government has done nothing to change that. In the absence of guiding laws, policies or tax incentives, it is entirely up to each of us to reduce our carbon footprint.

THE VISION FOR 2020
New Zealand will be the world's first carbon-neutral country, by following the logic of sustainability, which is the key principle of the Resource Management Act. The principles of sustainability will guide all sectors of society – central and local government, business, agriculture, transport and consumers.

STRATEGIES TO ACHIEVE THE VISION
As the government is lacking leadership, our motto should be: 'When ordinary people lead, the leaders will follow'. With respect to carbon neutrality this means leading by example as well as pushing for effective legislation. The government must pass the Resource Management (Climate Protection) Amendment Bill 2006, a private members bill. We need to develop national environmental standards and/or a national policy statement for achieving carbon neutrality. Underpinning all these strategies, there needs to be a commitment by each of us to establish a true and honest dialogue between government and civil society.

INTRODUCTION

Reducing our carbon footprint is a moral issue for every single New Zealander. It is also a public moral issue because our nation's economic prosperity is largely based on its carbon burdens being carried by poorer nations and future generations. If this is to change, public morality has to change. A key component of promoting public morality is the law for, without laws, neither government, society, nor individuals can be held accountable. The way climate change issues are legislated – or not legislated – is an indicator of how seriously carbon neutrality is being sought. Clearly, in New Zealand, current legislation is insufficient. Worse, our present climate change policies mean that we are destined to not even reach our rather modest Kyoto target.[1]

How then can legislation be improved? This chapter examines recent developments with respect to legislation controlling greenhouse gases. The Resource Management Act 1991 (RMA or the Act) had been designed with greenhouse gases in mind. It allowed for local authorities to consider the effects of an activity on climate change when assessing individual resource consent applications or when writing local plans.

However, these powers were taken away by the Resource Management (Energy and Climate Change) Amendment Act 2004 (the 2004 Act), which removed greenhouse gases from the RMA's scope. The 2004 Act was intended to clear the way for a carbon charge as a national economic instrument and centrepiece of the government's 2002 proposed climate change package.[2] A carbon charge would have created disincentives for the burning of fossil fuels and would have made renewable energy (wind, solar, biomass) more cost-effective. To date, no such carbon charge or any other national instrument has been introduced. In December 2005, the government specifically dropped its proposal for a carbon tax. Effectively, the RMA had lost its capacity as a regulatory framework for controlling greenhouse gases.

Instead of hard law, we are left with soft policies. Under the title '*New Zealand's Climate Change Solutions*',[3] the government recently released a set of programmes promoting energy efficiency, waste reduction, public transport and tree planting. It also announced a carbon-trading scheme that includes forestry and land management, but is vague on who will be involved and how carbon emissions should be priced. While carbon trading and incentives for energy saving are important, they are market-based economic instruments that can work only within a regulatory framework.

To restore this initial role of the RMA, Jeanette Fitzsimons, Co-Leader of the Green Party introduced a private members bill to the House: The Resource Management (Climate Protection) Amendment Bill 2006 (the Bill). The purpose of this bill is to once again empower local authorities to consider the effects of greenhouse gases, and in particular CO_2, on climate change when writing local plans and granting air discharge consents.

This chapter aims to show that the Bill is indeed needed to remedy a large gap in current climate change policies. If accompanied by national environmental standards or a national policy statement, local authorities would be enabled to enforce and monitor emission reductions so that New Zealand could eventually achieve carbon neutrality.

NEW ZEALAND'S INTERNATIONAL OBLIGATIONS

Under Article 2 of the 1992 UN Framework Convention on Climate Change,[4] New Zealand has undertaken to stabilise and, where possible, reduce emissions of greenhouse gases to 'a level that would prevent dangerous anthropogenic interference with the climate system'. Part of this commitment is to formulate and implement[5] national and regional measures to combat climate change.[6] Under the 1997 Kyoto Protocol, New Zealand is required to maintain a 0 per cent increase in emissions from 1990 levels in the commitment period 2008–2012. Further, by 2005, as a country that ratified the Kyoto agreement, we are bound to have made demonstrable progress in meeting such commitments.[7]

New Zealand ratified[8] the Kyoto Protocol through the Climate Change Response Act 2002. However, this act made only some institutional adjustments such as giving powers to the Minister of Finance to manage emission trading. The actual implementation of Kyoto obligations was left to further consultation and policy development.[9] One of the key considerations here had been the introduction of a carbon tax. However, this option was abandoned in late 2005. To date, no other policies to replace a carbon tax have been implemented, leaving us without any legal or policy instruments to make the necessary reductions in our greenhouse gas emissions.

The current uncertainty in New Zealand mirrors the international picture. There is uncertainty about the future of international climate change obligations. For example, how will non-compliance with Kyoto be dealt with in real terms? Will the USA or Australia become part of a post-Kyoto regime? Will China, India and Brazil accept emission targets? Will there even be a regime with any emission targets at all? For a post-2012 deal to become a reality, attitudes of the main actors would have to change dramatically. What we are seeing at present resembles a trench war where no country makes a move. The tensions are not only between USA and the rest of the world, but equally between the rich industrialised nations and the fast-growing economies of China, India and Brazil.

At the heart of the battle is an equity issue. The over-developed nations count for 80 per cent of global greenhouse gas emissions. Unless they demonstrate real leadership in reducing emissions, they cannot expect developing nations to become active partners. For New Zealand, proud of its reputation as a responsible world citizen, it is important to lead and this opportunity is now emerging. To overcome the deadlock in the negotiations, the European Union has recently committed to reduce emissions by 20 per cent (relative to 1990 levels) by 2020, encouraging others to follow. The history of climate negotiations has shown that the combined force of the EU, some allied countries and a large number of developing countries is necessary for any progress. At present, New Zealand and Canada are the only allied countries that could help the EU to lead the two sides out of the trenches. Small as New Zealand may be, its political weight right now is significant. New Zealand must lead now or we will continue to be part of the problem rather than becoming part of the solution.

For New Zealanders this means increasing the pressure on the government and it also means that the government needs to use the instruments it already has available. The key instrument is the law. Through law, mere ambitions become enforceable and emission targets can be monitored. Legal measures are essential to achieving carbon neutrality.

CONTENT AND STRUCTURE OF THE RESOURCE MANAGEMENT ACT 1991
Most New Zealanders have heard of the RMA. We are all affected by it. The RMA is the main piece of legislation that sets out how we should manage our environment. When the RMA came into force in 1991, it signalled a new approach to environmental law not just in New Zealand, but worldwide, by making it a legal imperative that the management of all physical and natural resources be sustainable.[10]

One possible drawback of the RMA is its use of the term 'management'. Sustainable management, as used in the RMA, is narrower in focus than the broader, and preferable concept of 'sustainable development', famously described by the United Nations Commission on Environment and Development (the so-called 'Brundtland report') as: 'development which [meets] the needs of the present without compromising the ability of future generations to meet their own needs.' Such a term encompasses social, economic and environmental sustainability in a global sense.

The broader considerations of the social inequities and global redistribution of wealth were intentionally excluded by the choice not to adopt the 'development' term.[11] The narrower concept of sustainable management focuses only on the *environmental* effects of activities.[12] Thus, while social and economic factors are relevant to a consideration of whether an action is sustainable under New Zealand law, they are so only in a restricted sense, subject to ecological considerations.[13]

Section 2 of the RMA defines the term 'contaminant' to include: 'any substance (including gases ...)... that either by itself or in combination with the same, similar, or other substances, energy, or heat ... when discharged onto or into land or into air, changes or is likely to change the physical, chemical, or biological condition of the land or air onto or into which it is discharged.' Given the international consensus on the causes of climate change, this definition is certainly wide enough to include greenhouse gases such as CO_2 and methane.

Part II of the Act requires consideration of listed matters of national importance,[14] the Treaty of Waitangi,[15] and other matters including efficient use and development of natural and physical resources, intrinsic values of ecosystems and the maintenance of the quality of the environment.[16] These sections are all preceded by the words 'in achieving the purpose of this Act', meaning that the sustainable management principle is intended to be paramount.[17]

Part III of the RMA concerns the control of the use of land, air, water and coastal areas. In the context of considering climate change, section 15 controls the discharge of contaminants into the air, as well as into water and land. As mentioned earlier, the definition for contaminants is broad enough to include greenhouse gases.

Part IV of the Act divides responsibilities between three sectors: central government, regional councils and territorial authorities. Central government's role is largely vested in the Minister for the Environment, who is empowered to regulate national environmental standards and can recommend the formulation of a national policy statement. The latter provides national policy guidance for matters that are considered to be of environmental importance. (These important tools are discussed more fully later in the chapter.) This way they help local government decide how competing national benefits and local costs should be balanced. In addition to setting national environmental standards or a national policy statement, the Minister also has the power to intervene or call in applications for a resource consent and for changes of regional plans.[18]

Regional councils play a pivotal role in the management of greenhouse gases as they are required to formulate regional policy statements.[19] These set the basic direction of environmental management in the region. They must be consistent with, and give effect to, a national policy statement, if there is one. In addition, regional councils can develop regional plans containing regulations to enforce their objectives.[20]

Part V of the Act provides details of the various types of policy and planning instruments. It describes, in particular, the purpose and content of regional policy statements and plans. The purpose of the statements is to: 'achieve the purpose of the Act by providing an overview of the resource management issues of the region and policies and methods to achieve integrated management of the natural and physical resources of the whole region'.[21]

All regional councils must have policy statements that give an overview of relevant resource management issues and prescribe the process by which decisions under the Act will be made. Regional plans are more specific in focus, their purpose being to enable regional councils to actually achieve the purpose of the Act.[21] Among other factors, plans must state the particular activity to be dealt with (for example, transport), further the objectives and policies relating to the plan, the methods to be adopted to achieve those ends and the reason for their adoption.[22] Unlike policy statements, regional plans may create rules prohibiting or regulating activities, which have the effect of a regulation under the Act.[23]

Part V also contains details relating to national environmental standards and national policy statements, both of which are relevant to implementing a climate change policy under the RMA.

THE RESOURCE MANAGEMENT ACT 1991 AS AMENDED IN 2004
With respect to climate change, the RMA provides important guidance. As mentioned in the previous section, the RMA's definition is certainly wide enough to include greenhouse gases such as CO_2 and methane, given the international consensus on the causes of climate change.

Section 5(2) defines sustainable management as: 'Managing the use, development, and protection of natural and physical resources in a way, or at a rate, which enables peoples and communities to provide for their social, economic, and cultural wellbeing and for their health and safety while:

- Sustaining the potential of natural and physical resources (excluding minerals) to meet the reasonable foreseeable needs of future generations; and
- Safeguarding the life-supporting capacity of air, water, soil and ecosystems; and
- Avoiding, remedying, or mitigating any adverse effects of activities on the environment.'

Section 7 pertains to what is considered 'appropriate' development, listing, for example, (f) 'Maintenance and enhancement of the quality of the environment'; (i) and (j) 'The effects of climate change and the benefits to be derived from the use and development of renewable energy'.

These subsections (i and j) were explicitly added by the 2004 amendment of the Act. However, section 7 together with section 104E suggests that the effects on climate change can be considered only with respect to the development of renewable energy. Section 104E says: 'When considering an application ... a consent authority must not have regard to the effects of such a discharge on climate change, except to the extent that the use and development of renewable energy enables a reduction in the discharge into air of greenhouse gases, ...'

In practice, the RMA is administered by local authorities. They issue consents to businesses, developers and individuals whose activities will impact on the physical environment of the area. This includes activities like building, mining and constructing manufacturing plants. Given the 2004 amendment, to what extent can local authorities consider climate change?

This question has recently been dealt with by the Environment Court and the High Court, when deciding on a discharge consent for a coal-fired power station. In 2005, the Northland Regional Council granted consent to Mighty River Power for the operation of the Marsden B power station near Whangarei. Greenpeace Aotearoa New Zealand appealed against this decision, arguing that coal has an effect on climate change and should be considered under section 104E of the RMA.[24]

The Environment Court rejected this argument. It decided that 'section 104E specifically prohibits [consent authorities] from considering the effects of climate change ...'.[25] Therefore, the effects of the discharges from a coal-fired power station on global warming could not be taken into account. Greenpeace appealed and, in October 2006, the High Court reversed the Environment Court's decision. According to the High Court, section 104E also applies to a non-renewable energy project in so far as it may be beneficial to greenhouse gas reduction. Consent authorities should be able to balance the benefits of a hypothetical renewable energy project against the actual application concerning a non-renewable energy project.[26]

In March 2007, Mighty River Power abandoned its plans for Marsden B. However, a few months later, Genesis Energy, a state-owned enterprise and operator of the Huntly Power Station, filed new legal proceedings in a bid to reverse the High Court ruling. Genesis claimed that the ruling 'took most by surprise in reversing an amendment to the RMA which was designed to remove greenhouse gas issues from the individual consent process to the national level'.[27] Quite obviously, energy companies fear the full impact of the law.

The wider implications of the High Court's ruling are not yet clear. On the one hand, the objective of the 2004 Act was to deal with climate change on a nationally uniform basis. The Court's interpretation of section 104E would meet this objective.[28] On the other hand, in removing the ability for local authorities to control greenhouse gases, the government was keeping all its policy options open.[29] Clearly, the court's interpretation is at odds with the government's complete reliance on market-based economic instruments. With its announcement of a sector-wide emission trading system, the government is further moving away from the RMA. A regime of tradeable permits could be established for any sector and any number of carbon emitters and could operate independently of the RMA.

The government is at a crossroads now. If it continues to rely on economic instruments, market forces will decide what is possible and what is not. Although policies require a regulatory approach, the same can be said for economic instruments. In fact, economic instruments and regulatory measures are not mutually exclusive: they can and should work hand-in-hand, but a suitable overall regulatory framework is necessary.

With the RMA, New Zealand has, in its wisdom, created the appropriate legislative framework. The RMA allows for enforceability and predictability of measures, but also for flexibility. Flexibility is provided by the fact that the Minister for the Environment can prescribe national environmental standards.[30] Maximum emission standards for greenhouse gases, for example, are relatively easy to establish and can be changed to reflect ongoing practical or technological advances.

THE IMPORTANCE OF NATIONAL ENVIRONMENTAL STANDARDS AND POLICY STATEMENTS

National environmental standards are tools used to set nationwide standards for the state of a natural resource. For example, 14 standards for the prevention of toxic emissions and the protection of air quality were introduced in October 2004. National environmental standards come in the form of regulations issued under sections 43 and 44 of the RMA and apply nationally. This means that each regional, city or district council must enforce the same standard, securing a consistent approach and decision-making process throughout New Zealand. (In some circumstances, local authorities can impose stricter standards.) They may concern contaminants, water quality, air quality and soil quality, in relation to discharge of contaminants and provide standards, methods or requirements for monitoring.[31]

The procedures for establishing the standards were reformed in 2003. Under the new procedures, the Minister for the Environment must notify the public of the proposed subject matter and show that the reasons for considering the regulations are consistent with the purpose of the RMA. The public must then be given adequate time and opportunity to comment. Based on these comments, a report with recommendations is then written and publicly notified. At the end of these consultative procedures, the regulations can then be promulgated.[32]

The first formal regulations prescribing national environmental standards were issued in 2004. These regulations relate to certain air pollutants, dioxins and other toxics,[33] but not to greenhouse gases. If the 2006 Resource Management (Climate Change) Amendment Bill is passed, then standards and rules for greenhouse gas emissions could be drawn up. Their practical importance is that they could prohibit or restrict certain discharge activities. And they would apply to resource consent applications and the drafting of regional and district plans.[34]

The existing regulations cover a number of specific activities – for example, the burning of waste in landfills, the operation of incinerators near schools and hospitals, and the installation of wood burners. Schedule One to the regulations sets out the ambient air quality standards for contaminants (excluding greenhouse gases). The only controls of greenhouse gases are imposed in relation to landfills.[35] The current state of legal affairs is not without irony. While a system of air quality standards has been established, it protects people from certain toxins with an immediate effect on health, but almost completely ignores the biggest threat – climate change.

Another tool is a national policy statement. This would provide national policy guidance for matters that are considered to be of environmental importance, for example, the coastal environment. The purpose of a national policy statement is to state objectives and policies for matters of national significance relevant to achieving the purpose of the RMA.[36] For the decision whether one is appropriate, the Minister for the Environment has to consider a number of factors, particularly aspects of the national and global environment.[37] Reducing New Zealand's carbon emissions would be a classic example. One reason for the inclusion of national policy statements in the RMA was that they allow the government to be proactive, setting down broad objectives from the start, rather than having to object at the consent stage.[38]

Apart from the New Zealand Coastal Policy Statement, no national policy statement has been issued to date.[39] A major reason for their absence has been that the drafting procedures are very complicated.

If the Minister considers it desirable to issue a national policy statement, either a standard procedure or a more informal alternative procedure can be followed.[40] Either way, the Minister must seek and consider comments from relevant groups. If the standard procedure if followed, a report with recommendations is then referred to the Minister for consideration. The Minister must then prepare a proposed national policy statement and appoint a board of inquiry to inquire into and report on the proposed statement. The board must invite submissions by a closing date (sections 47.48). The board's report and recommendations are then presented to the Minister who has the discretion to make changes (section 52) before making a recommendation to the Governor-General for an Order in Council. Finally, the national policy statement is published and public notice is given of its issue.

A national policy statement cannot be directly enforced; rather, regional policy statements, and regional or district plans must be consistent with it. When a national policy statement is issued, local authorities are required by the RMA to make all necessary changes to their existing policy statements or plans so as to remove inconsistencies and conflicts. They must also take extra steps necessary to implement the national policy statement, and give public notice if they intend to do this.[41] Where there is a dispute over whether there are in fact inconsistencies between a national policy statement and a regional policy statement or plan, or whether the latter needs to be changed, the Environment Court has jurisdiction to decide the point.[42]

In 1998, the complexity of this process was a subject of concern for the Minister for the Environment's Reference Group (which reviewed the operation of the RMA). The group pointed out that the only existing national policy statement, the New Zealand Coastal Policy Statement, took over two years to prepare, not counting draft preparation time.[43] The process was seen as excessive and as a major reason why, to that point, no further national policy statements had been prepared.[44]

As a result of this review, the 2006 Bill contains a substantial change to the process of formulating a national policy statement. If passed, provisions relating to the mentioned board of inquiry would be repealed.[45] Instead, a new section 46(3) would require the Minister to establish 'a process that gives the public adequate time and opportunity to comment on the subject matter of the proposed [national policy statement, NPS, and ...] an advisory committee to provide advice to the Minister on the NPS and the comments made on that statement'.[46]

Having an advisory committee rather than a board would allow the Minister more scope to change the recommendations offered. It would also considerably reduce the requirement for public consultation, with the likely result that the second stage of submissions would be removed.

The amendment would also remove the need for a local authority to give public notice regarding its decision on how to implement the national policy statement, and the reasons for such a decision.[47] While the amendment would decrease the opportunity for public participation in the formulation of a national policy statement, the length of time involved in formulating the Coastal Policy Statement has clearly acted as a disincentive to the creation of a national policy statement on climate change. Reforming the process is certainly desirable considering the urgency of reducing our carbon emissions.

Stratford recommendation 1995
The issue of providing a national policy statement was discussed at some length in the so-called 'Stratford Inquiry'. A number of submissions urged a board of inquiry to recommend a national policy statement under section 148(c).[48] After examination of the arguments presented for both sides, the board of inquiry decided to recommend an NPS on CO_2 emissions.[49]

The two main advocates for an NPS were Greenpeace and Taranaki Regional Council, the local authority controlling the area in which a proposed power station was to be built. Since the Taranaki region has a significant number of carbon emitters, and therefore considerable knowledge of the concerns involved, the board considered the submissions of Greenpeace and Taranaki Regional Council to have great weight.[50] A major reason influencing the board's decision to recommend a national policy statement was the wish to avoid procedural costs. It acknowledged that in the near future, there may be more applications to build new power stations and, since the same national interest would apply, the applications would be likely candidates for interventions of the Minister (the earlier mentioned call-in power), and another inquiry would need to be instigated.[51] The board considered that the existence of a national policy statement would eliminate the need for a call-in procedure and the possibility of inconsistent decisions being made.[52]

A national policy statement was also seen as a way of providing guidance to local government when it is placed in the 'invidious position' of having to take into account factors that are not local, but are matters of national and, indeed, international, policy in their everyday decisions.[53]

In the course of the inquiry, the (then) Secretary for the Environment voiced opposition to a national policy statement:

> Because of the global nature of the problem, and the national nature of the commitments [a national policy statement] does not seem the most efficient mechanism to implement policy. National climate change commitments on greenhouse gas emissions do not lend themselves to being specifically subdivided for regional application.[54]

The board rejected this view and recommended that the process of developing a national policy statement be initiated. It also made some comments as to the possible content of the statement. It stated that any national policy statement on energy or climate change should contain policies governing discharges of CO_2 from all anthropogenic sources in New Zealand and should be designed to implement New Zealand's international obligations to the global environment.[55]

On receipt of the recommendation, the Minister stated that he would defer any decision regarding a national policy statement until such time as he had undertaken further discussions with colleagues. Consideration of the advice was effectively left to the Working Group on CO_2,[56] whose attitude to using the RMA mechanisms was hostile.[57]

Central government has repeatedly expressed concerns of inconsistency between regions as an argument against using the RMA.[58] Yet none of these concerns have ever been substantiated, nor has central government, to any greater extent, considered the critical role of the RMA and of local government for effective reductions.[59] As we have seen, both national environmental standards and national policy statements would ensure national consistency among local authorities. They would also avoid the considerable expense and uncertainty engendered in Ministerial call-ins.[60] Whether a more technical national environmental standard or a more comprehensive national policy statement is preferable may be a matter of practicality, according to the Bill's author, Jeanette Fitzsimons:

> An NES (National Environmental Standard) as provided for in the Act would be more effective. That could then set standards for offsets or mitigation for carbon emissions which regional councils would find easy to apply. An NPS would be high level policy guidance, but what they need to know is how to decide how much forest should be planted to offset how much carbon, and what conditions should be put on that forest. Many submitters supported adding a requirement for an NES to the bill ... I have always opposed (and when I chaired the select committee, successfully) short-cutting the NPS process. It is meant to get wide public buy-in for fundamental policy and needs proper consultation. It should not be changed with every change of govt. An NES could be achieved much faster, which is another reason for preferring it.[61]

Given the urgent need to provide meaningful guidance to local authorities (if they are again empowered to consider CO_2 emissions), national environmental standards may indeed be the better option. A national policy statement should certainly also be developed as the basis for meaningful long-term policy responses to climate change.

Environmental Defence Society v Auckland Regional Council 2002[62]
Seven years after the Stratford recommendations of a national policy

statement to address with carbon emissions, the Environment Court had the opportunity to decide on the applicability of the RMA for greenhouse gases. The Environmental Defence Society appealed against a consent issued for a gas-powered power generation plant, known as 'Otahuhu C', for the discharge of 1.2 million tonnes of CO_2 per year.

The appeal was dismissed by the judge, J. Whiting, on the grounds that the national and international consequences could not be adequately assessed and that the New Zealand government preferred to address greenhouse gases in a different way.[63]

The logic of this is questionable, of course. The Stratford board of inquiry had roundly rejected such reasoning: 'To do so would imply that as the world's CO_2 is composed of a great number of small emissions, the effect of any one of them could be discounted.'[64]

To further quote Whiting:

> The government has already signalled that it does not see RMA controls and the mechanisms as being cost effective for managing greenhouse emissions. Climate change is an international issue and should therefore be dealt with consistently on a national level. The RMA consenting and planning process means that there will always be a risk of inconsistent treatment and costs of implementing and managing requirements for different regions.[65]

These comments only reinforce the need for national environmental standards or a national policy statement for climate change.

REVERSING THE TREND

The 2006 Bill would reinstall the law as it was before the 2004 Amendment Act. The use of the RMA with national environmental standards or a national policy statement would enable local authorities to develop much more effective policy instruments. To date, local government has received little or no encouragement from central government to address climate change issues.

The Bill would, for example, allow the Auckland Regional Council to extend the current provisions in the Auckland Regional Policy Statement[66] relating to the effect of greenhouse gases, air discharges, waste management and the role of energy, to include a regulatory role embodied in regional plans. Regulations could control greenhouse gases in much the same way as regional plans currently include provisions under the 2004 regulations, discussed above.

Combining the management of both emissions and land-use controls in the same authorities also provides opportunities for synergies. For example, councils frequently include planting or revegetation as a condition of a land-use resource consent, probably partly to offset increased transport emissions, if these are likely to arise. Explicitly empowering or directing local government to take some responsibility for the management of greenhouse gases provides opportunities for this global issue to be addressed through all levels of society. The government recognises that there is need for an increased level of public awareness of global warming, if the country is to respond effectively to climate change. For many people, one of their most significant interfaces with any sort of government is through their contact with local authorities, and in particular for resource management or planning purposes. If these local authorities deal with global warming as part of their functions, their ability to inform the public about the relationship between their own actions and climate change will be greater.

With the RMA, New Zealand has adopted one of the most efficient environmental tools in the world. To not use it with respect to climate change makes no sense considering the urgency of action. It contains all the instruments for enforcing and monitoring carbon reductions.

Notably, the government's website on climate change solutions spells out a vision: 'New Zealand can become a global brand leader – the world's first clean, green and sustainable economy'. It also claims that New Zealand could become:

'Carbon neutral in the electricity sector by 2025;
Carbon neutral in the stationary energy sector by 2030;
Carbon neutral in the transport sector by 2040;
Carbon neutral in the total energy sector by 2040.'[67]

These targets are laudable, but none will ever be achieved without making carbon reductions enforceable. It is only through the combined forces of RMA, national environmental standards, national policy statements and local authorities that carbon neutrality will be achieved. The ultimate commitment, however, is personal. Nothing will ever change unless we change individually. We cannot rely on others. Perhaps a good start is asking the question: How will I explain to my grandchildren what I have done about climate change?

ENDNOTES

1 Bosselmann, K. 2006. Achieving the goal and missing the target: New Zealand's implementation of the Kyoto Protocol *Macquarie Journal for International and Comparative Environmental Law 2*, 75–106.
2 For history and details of the government's 2002 'Preferred Policy Package on Climate Change' see Bosselmann, J., Fuller, J. and Salinger, J. 2002. Climate Change in New Zealand: Scientific and Legal Assessments, *NZCEL Monograph Series 2*, 51–109. Auckland: New Zealand Centre for Environmental Law.
3 To find out more about the programmes, visit http://www.climatechange.govt.nz (accessed 1 Oct 2007).
4 To find out more about the United Nations Framework Convention on Climate Change 1992, visit http://www.mfe.govt.nz/issues/climate/international/unfccc.html (accessed 1 Oct 2007).
5 Article 4(2) of the United Nations Framework Convention on Climate Change 1992.
6 Bosselmann, K. et al. 2002. p. 51.
7 Bosselmann, K. et al. 2002. p. 51.
8 Ratification is a parliamentary act that incorporates international law into domestic law.
9 Cabinet paper, September 2001.
10 Nolan, D. 2005. *Environmental Law and Resource Management Law*. 3rd edn., para. 1.2. Wellington: LexisNexis. See also Palmer, K. 2002. Origins and Guiding Ideas of Environmental Law. In: Bosselmann, K. and Grinlinton, D. (Eds). Environmental Law for a Sustainable Society. *NZCEL Monograph Series Vol. 2*, p. 3, at 17. Auckland: New Zealand Centre for Environmental Law.
11 Nolan, D. 2005. para. 3.2.
12 Ministry for the Environment.1991. *Environment Update Series, Information sheet No. 6*, (December 1991) cited in Nolan, D. 2005, para. 3.8.
13 Nolan, D. 2005. para. 3.10.
14 RMA, s6.
15 RMA, s8.
16 RMA, s7.
17 See *Falkner v Gisborne District Council* [1995] 3 NZLR 18–20, cited in Nolan, D. 2005. para. 3.15.
18 RMA, ss140-150AA (as amended 2004).
19 RMA, Second Schedule
20 RMA, s63.
21 RMA, s63.
22 RMA, s67.
23 RMA, s68(2).
24 RMA, s68(2).
25 *Greenpeace New Zealand Inc v Northland Regional Council* (Environment Court, Auckland, A94/06, 11 July 2006, Judge Newhook) para. 4.
26 Ibid., para. 51; see also Warnock, C. 2006. Greenhouse gases and climate change, *Resource Management Bulletin*, 191–192.
27 Genesis Energy, Press Release, 23 May 2007. It should be noted that Genesis is aiming for the development of carbon storage technologies at its Huntly facility.
28 Warnock, C. 2006. p. 193.

29 Bosselmann, K. et al. 2002.
30 Under RMA, s 43.
31 RMA, s43.
32 RMA, ss.43A-G (as amended 2003).
33 Resource Management (National Environmental Standard Regulations to Certain Air Pollutants, Dioxins, and Other Toxics) Regulations 2004 (SR 2004/309).
34 Nolan, D. 2005. para. 3.74
35 Resource Management (National Environmental Standard Regulations to Certain Air Pollutants, Dioxins, and Other Toxics) Regulations 2004 (SR 2004/309). Regulations 25–27.
36 RMA, ss45(1), 60(1).
37 RMA, s63(1); see also Nolan, D. 2005. para. 10.34
38 Ministry for the Environment. 1998. *Resource Management Act 1991: Report of the Minister for the Environment's Reference Group* Wellington: Ministry for the Environment.
39 The New Zealand Coastal Policy Statement is mandatory and has been prepared and recommended by the Minister of Conservation; RMA, ss46–52.
40 The alternative procedure is like the procedure for preparing national environmental standards. However, it may not be used for matters to be inserted in regional and district plans without further public notice; RMA, ss46A,55.
41 RMA, s55; Nolan, D. 2005. para. 3.77.
42 RMA, s82.
43 Ministry of the Environment. 1998. p. 43.
44 Ministry of the Environment. 1998. p. 43.
45 Resource Management Amendment Bill, v.
46 Resource Management Amendment Bill cl. 20.
47 Resource Management Amendment Bill cl 21, amending RMA, s55.
48 Williams, D. QC. 1995. *Proposed Taranaki power station; air discharge effects: Report and recommendation of the Board of Inquiry, pursuant to Section 148 of the Resource Management Act* (the 'Stratford Inquiry'), 1995, p. 209.
49 Williams, D. 1995. p. 235.
50 Williams, D. 1995. p. 212.
51 Williams, D. 1995. p. 220.
52 Williams, D. 1995. p. 220–221.
53 Williams, D. 1995. p. 220. See also 211.
54 Williams, D. 1995. p. 216.
55 Williams, D. 1995. p. 222.
56 Working Group on CO2 Policy 'Climate Change and CO2 Policy', 1996.
57 See Bosselmann, K. et al. 2002. p. 57.
58 Bosselmann, K. et al. 2002. p. 57, 102, 114.
59 Bosselmann, K. et al. 2002. p. 54.
60 Bosselmann, K. et al. 2002. p. 106.
61 Jeanette Fitzsimons, email, 15 Feb 2007.
62 [2002] NZRMA 492.
63 [2002] NZRMA para. 88.
64 Williams, D. QC. 1995. para. 7.103.
65 [2002] NZRMA 492. para. 18 (v).
66 The policy statement can be ordered through http://www.arc.govt.nz/arc/publications/regional-policy-and-plans/ak-rps.cfm (accessed 1 Oct 2007).
67 See http://www.climatechange.govt.nz/sustainable-nz/sustainable-future.shtml (accessed 1 Oct 2007).

Political activism
and carbon neutrality

Jennifer Curtin and Anita Lacey

Jennifer Curtin is a Senior Lecturer in the Department of Political Studies at the University of Auckland where she teaches comparative public policy, models of policy-making and critical approaches to the politics of policy change. The activism and influence associated with environmental movements and networks feature in her teaching. Jennifer has published widely on women's political activism and representations of gender in politics and policy and cross-national comparisons of trade union activism. She is currently working on a research project that explores gender and constructions of political leadership.

Anita Lacey is a Lecturer in international relations in the Department of Political Studies at the University of Auckland. Anita researches and publishes in the following areas: the governing role of international organisations; poverty reduction and non-governmental organisations; gender and human security; and protest networks, including transnational environmental activist networks and the global justice movement and autonomous protest spaces. She teaches in the areas of international relations and global institutional governance, security studies and comparative foreign policy. All aspects of her work are informed by a goal of drawing together environmental, gender and social justice.

THE CURRENT SITUATION
Initiatives for carbon neutrality are being generated and pursued by a wide range of political actors. Sixteen cities and municipal councils around Aotearoa/New Zealand have implemented practices that support carbon-neutral goals as part of the international Cities for Climate Protection Plan. The Labour Party-led government has put in place carbon-neutral plans for all of its ministries and departments. Last, but certainly not least, environmental activists are demonstrating that by creating on-line networks and acting collectively they can effect change both directly and indirectly.

THE VISION FOR 2020
In 2020, we will all be fully aware of the importance of protecting the environment and will be politically engaged in ensuring Aotearoa/New Zealand is, and will continue to be, a carbon-neutral society. While different parties may have different strategies for achieving this, there will be a broad political desire in parliament and at the grassroots level to maintain carbon neutrality and ensure environmental protection and regeneration beyond 2020. A new public policy will require a carbon-neutral impact statement before it can be passed into law and implemented. And we as citizens will be participating in on-line and 'real' political activism to ensure governments of all persuasions, in Aotearoa/New Zealand and abroad, maintain carbon-neutral positions and practices.

STRATEGIES TO ACHIEVE THE VISION
We need to build on our long history of transnational and local environmental activism that already has led to significant policy change in this country. The urgency of carbon neutrality requires us to think and act strategically about how we can collectively have an impact on our government. This means we should be voting, lobbying and initiating petitions so that our politicians see how important this issue is to us. We should also be exchanging ideas and practical carbon-neutral strategies and, as much as possible, living these strategies in our daily lives. We need to develop a plethora of networked organisations, local and transnational, that will provide politicians, bureaucrats and other policy makers with up-to-date research to help them to generate progressive policies around carbon neutrality and other important environmental issues.

INTRODUCTION

Politics matters. Yes, it is important that we change our own individual behaviour and become more responsible in choosing how we travel, what we consume and how we purchase our energy. But we also need to think and act politically, and collectively, in order to revolutionise our communities' and governments' thinking on carbon neutrality. There are many local activist groups and individuals who are already demanding that governments at the local, national and international levels address environmental issues generally and carbon neutrality in particular. Their work has been very effective in ensuring that the current Labour-led government has taken the first steps towards establishing a carbon-neutral public service. But there is still much more we can do. This chapter explores a number of other possible strategies that we could all be involved in to ensure that current and future governments see carbon neutrality as a political and policy issue that we, as citizens of Aotearoa/New Zealand, desperately want addressed.

To guarantee that the general public and governments view carbon neutrality as an ongoing critical issue, we must be vigilant in ensuring that our political demands are constant, relentless and have a consistently high profile over the next decade. Politicians need to know that this is an issue that could lose them an election at any time. This is a difficult task. There are always many more policy 'problems' of concern to us as citizens than the government can deal with at any one time. So, in a sense, there is competition between interests and issues for a place on the policy-making agenda. In addition, it is well known that broad public interest in policy 'problems' can be fleeting.[1] The short attention span we have as citizens is, in part, fuelled by a media that focuses on headlines and sound bites rather than sustained investigative journalism.

Alongside this, we know that the proportion of people turning out to vote in national elections has been in decline over the past 10 years, not only in Aotearoa/New Zealand but in many countries around the globe. Many people feel politics is not relevant to their lives, that politicians are out of touch, and that political activism is not for them. Young people in particular appear disinterested in participating in traditional forms of political action such as joining political parties and interest groups, registering themselves on the electoral roll and then turning out to vote. These findings might suggest that mobilising people to act politically in order to achieve carbon neutrality is futile. However, the issues of ecological protection and carbon neutrality have real potential to transcend our current disinterest in politics. There are many different means by which we can demonstrate clearly to our national government, and to the world, that we are prepared to exert our collective political power in order to advance the cause of carbon neutrality. In the remainder of this chapter some radical ideas and actions around the fight for carbon neutrality are reviewed, and others are identified that are more mainstream. All of these require us as individuals and as members of networks, local or transnational, to engage directly with the state. Focusing on the possibilities of multiple sites and types of action enables us to imagine a political activism that supersedes traditional political identities and produces broad-based and continual pressure for a carbon-neutral Aotearoa/New Zealand.

POLITICAL ACTIVISM IN CONTEXT

Environmental issues are not new to the political arena in Aotearoa/New Zealand. The process of colonisation in the mid-1800s and with it the appropriation of land and resources, along with the introduction of foreign species, challenged both indigenous culture and ecology. Alongside Maori objections to this appropriation, an environmental consciousness emerged amongst some Pakeha, resulting in a variety of political strategies focused on native forest protection. This attention to nature preservation continued to feature as the lead environmental issue in Aotearoa/New Zealand for much of the first half of the 20th century.[2]

In the 1970s, with the arrival of the new environmental movement, a heightened, and more radical, activism emerged around natural resource exploitation and the global implications of ecological degradation. Within Aotearoa/New Zealand, a number of groups emerged including, for example, Ecology Action, which focused on issues of conservation and nuclear power. Similarly, the Campaign for Nuclear Disarmament, established in the 1960s, regained momentum and was successful in removing nuclear power from the energy agenda, while the Royal New Zealand Forest and Bird Protection Society spearheaded a massive public protest to save Lake Manapouri.[3] These groups, like other new social movements of the time, used unconventional forms of political participation including protests and demonstrations, sit-ins, alternative publications and mass petitions, which were highly effective in drawing attention to environmental issues of concern.

Transnational activism also became more explicit during this time, with the Aotearoa/New Zealand branch of Friends of the Earth established in 1974, four years after its formation in the USA. Greenpeace Aotearoa New Zealand (hereafter, Greenpeace) was also established in 1974, stimulated partly by connections between the Canadian founders of Greenpeace and New Zealand anti-nuclear activists. Over the course of the 1970s and 1980s, conservation and nature preservation tended to dominate the environmental agenda in Aotearoa/New Zealand and culminated in the institutionalisation and state funding of conservation through the establishment of the Department of Conservation in 1987. Carbon neutrality did not appear on the political agenda in any real sense until the early 1990s.

MAKING OUR VOTES COUNT

The Winter 2007 issue of Greenpeace's magazine features in its headline 'People Power' and includes photos of protesters carrying placards stating 'I love clean energy and I vote'. In her editorial comment, Bunny McDiarmid reflects on the power of protest and the vote in achieving the nuclear-free legislation 20 years ago. Voting is the most basic, but collectively one of the most powerful, tools we have as citizens to influence government and its policy agendas and to bring about change.

So, it is not surprising that while grassroots decision-making and non-violent direct action were key themes among environmental activist groups in the 1970s, the focus also turned to mainstream politics, with the creation of the Values Party in May 1972. Indeed, the Values Party is recognised as the world's first national Green party; its members espoused radical ideas of zero economic growth and zero population growth alongside broader ecological goals. In 1975, they won 5.2 per cent of the vote, which under an MMP system would have resulted in parliamentary representation. But it was not until the arrival of MMP in 1996 that a Green presence in parliament became a reality, and in 2005, the Green Party held six seats in parliament.[4]

Our political influence as individual citizens increased with the advent of MMP, because the new system ensures that our votes are not wasted. In other words, the proportion of votes a party receives is what determines the proportion of seats it wins in parliament. In addition, the electoral outcomes produced by the proportional representation element of the MMP system has forced both the major parties

(Labour and National) to pay more attention to the challenges posed by smaller parties.

In particular, the mere presence of the Green Party of Aotearoa as a viable option for us as voters can push other parties to consider adopting better environmental policies as part of their mainstream platform.[5] Why? Political parties are extremely competitive in their attempts to attract as many voters as possible. So, when a small party (for example, the Green Party) adopts a strong environmental position, it may take votes away from the major parties. This potential threat can stimulate a major party to mimic the smaller party in terms of policy. And if one of the larger parties does this, the nature of party competition means that the other major party, out of political necessity, is also likely adjust its own policy position. This is known as a policy 'contagion' effect and demonstrates that minor parties such as the Green Party can have considerable influence on public policy, even if they never even become part of a coalition government.

In Aotearoa/New Zealand, Labour is considered to be the closest major party to the Green Party of Aotearoa on the traditional political left–right spectrum. And in recent years, Labour has begun to address the issue of carbon emissions, with varying degrees of success. Labour ratified the Kyoto Protocol and in 2002 announced a package of policies that responded to climate change, including the now-defunct carbon tax. Currently, Labour has plans for Aotearoa/New Zealand to be the first carbon-neutral nation, whereby the 34 core public service departments will begin immediate work to achieve carbon neutrality from 2012.[6] The government is also promoting climate change awareness and local activism through a number of websites.[7] It is not possible to say whether such policy initiatives are a direct result of the presence or threat of the Green Party, but a contagion effect is a distinct possibility.

Of course, carbon neutrality is an issue of concern to more people than just Labour voters. The National Party clearly recognises this. Nick Smith, Shadow Minister for the Environment highlighted that only 6,822 votes saved the Green Party from dropping below the 5 per cent quota required to enter parliament. He went on to say that if National 'can convince just 1 in 20 of those Green voters that National is a better bet than the Greens, that alone would be enough, all other things being equal, for National to win in 2008.'[8] He spoke of convincing green-voting urban liberals to vote National, and as a response, the

National BlueGreens has been established as a policy interest group aimed at developing environment policies that have appeal to their core constituency. Perhaps another example of contagion?

So, it is clear that how we vote matters. And because we have two votes at each election, (one to choose our local representative and the other to choose our preferred party), we can vote for two different parties. This means we can use our party vote to ensure a minor party presence in Parliament. Thus, as voters, we have considerable 'power' to influence the advances made on the issue of carbon neutrality. As such, we must lobby parties to ensure they have a visible policy on carbon neutrality going into the next election; we must register ourselves on the electoral roll and turn out to vote on election day. And we need to continue to think strategically about how we use the two votes we have at each national election.

CARBON-NEUTRAL NETWORKS AND POLITICAL CHANGE

Many of us who have been involved in environmental activism recognise that voting in itself will not bring about the radical changes currently needed to achieve carbon neutrality. Protest action and the sharing of information between groups and networks are important political strategies in their own right, which can influence government. They can also raise our consciousness about these key issues, and empower us to make changes in our own lives. Those who have written about creating a green democracy have argued that more localised, grassroots and participatory networks are critical to the process of political participation and personal empowerment.[9]

While the word 'network' may seem rather vague, in the literature on transnational activism, the term refers to a non-hierarchical form of communication and interaction between activists.[10] The networks formed around carbon neutrality are both formal and informal, including parliamentary parties, such as the Green Party of Aotearoa, activists within and around these political parties, as well as loosely constituted activist groups such as those discussed below.[11] Moreover, many contemporary networks established around campaigns like carbon neutrality by 2020 are deliberately transient. In other words, they form around an issue, work intensely towards advancing the cause, but may be short-lived or go on to re-create themselves around a different issue or strategy for action.

This kind of political activism rarely involves a centralised and formalised membership. The example of the Karangahape Road Reclaim the Streets action in March 2007 is a case in point. This was one of hundreds of such days of actions worldwide calling for an end to dominant car culture and for concerted endeavours to bring about carbon neutrality and reverse climate change. The activists' attitude to achieving political change is summed up by an anonymous author on the Reclaim the Streets (UK) homepage: 'We are not a send-a-donation/get-the-mag/sit-in-your-armchair organisation. We are about getting involved and changing things through our own actions.'[12] For direct action activists, political change can be brought about by being involved and participating actively, rather than relying on formal party or other organisational membership as a conduit for change.

Sometimes the actions undertaken by such networks are illegal, increasing the value of participants remaining 'anonymous' and only loosely connected. The ongoing occupation by activists at Happy Valley near Westport in the South Island is an example of such action. Happy Valley is the proposed site of a Solid Energy opencast coalmine that threatens the valley's ecosystem and will produce over 12 million tonnes of CO_2. The Save Happy Valley Coalition has held demonstrations at the company's headquarters, blockaded Solid Energy coal trains and has staged an occupation of parts of the proposed site since January 2006. In a press statement at the beginning of the activists' continual occupation of the valley site, we can see the dual focus of carbon-neutral activism: a Save Happy Valley Coalition spokesperson urged people to put pressure on the government and also to join the (illegal) occupation of the Valley.[13] Anonymity is important to this form of protest action. Activists from the Coalition could be at risk of surveillance and arrest if their names and details were supplied for centralised membership list. For others, remaining an anonymous part of a wider collective is a political action in itself.

POLITICAL ACTIVISM, NEW TECHNOLOGY AND INTERCONNECTING NETWORKS

Effective political or activist networks and the 'foundations on which movements are built' are often dependent on the 'concrete linkages that derive from locality, shared experience, kinship, and the like…'.[14] However, technology has also become an important tool in facilitating the creation and growth of these networks. The potential of technology driven networks in fostering broader networks is well illustrated by the case of the transnational network for carbon neutrality. Rather than coming together at physically constituted conventions or conferences, activists share information and agendas for policy change via the Internet.

The advent of a transnational informal carbon-neutrality network, connected in the virtual world, means that a person active in the Karangahape Road action in Aotearoa/New Zealand or in Leeds EarthFirst! in Britain, or participating at the Camps for Climate Action in Europe or North America (discussed below) can be active in local campaigns while also participating in, contributing to and/or interacting in national and international actions. The transnational carbon-neutrality network has been promoted by the interaction of activists over the Internet. The network's community is not limited to a single geographical location. Instead, Barry Wellman and Milena Gulia argue that the virtual community, or EF! community in this case, is 'glocalized': 'Operating via the Net, virtual communities are glocalized. They are simultaneously more global and local, as worldwide connectivity and domestic matters intersect.'[15]

Therefore, one means of achieving carbon neutrality by 2020, we propose, is via a transnational environmental movement.[16] This can be a dynamic process of shared policies, tactics and information. Networks are intangible and activists can be part of a number of isolated or interconnecting networks. For example, an activist may learn of forthcoming Greenpeace or Reclaim the Streets campaigns through an on-line information page like Indymedia or of a new policy initiative on carbon-zero homes from the Toolkit for Climate Action.[17] The networks themselves are characterised by their non-hierarchical structure and the fluidity with which activists can move in and out of and between them. For example, in August 2006, Network for Climate Action, a decentralised network of autonomous groups and people, staged a Camp for Climate Action near the Drax coal power station in Selby, North Yorkshire. Over 600 people attended the 10-day camp, which focused on 'low-impact living, education, debate, networking, strategising, celebration and direct action'.[18] This year, the Camp for Climate Action took place near Heathrow Airport. The location was chosen to highlight the links between the British government's policies to expand the airport and the need to pursue holistic carbon-neutrality policies, targeting not only individual but also industry carbon emissions.[19] The Camps for Climate Action now also take place across North America and in continental Europe.[20]

In addition to its protest activity, the UK Camp for Climate Action helped to produce the Toolkit for Climate Action.[21] This Toolkit was developed by a wide cross-section of the carbon-neutral network in Britain, itself part of a broader transnational network, ranging from more highly institutionalised groups like Friends of the Earth and Greenpeace to loosely configured activist groups like EarthFirst!, SchNEWS and Radical Routes.

The carbon-neutral activist network used the Toolkit for Climate Action as a model and campaigning tool and the British government in turn developed the Home Information Pack. This mainstream initiative includes, for example, new energy-efficiency certificates for homes that 'provide a clear indication to buyers on the energy efficiency of a property – with ratings much like the energy efficiency labels provided on electrical goods'.[22] This example demonstrates the potential symbiotic relationship between types of political actions and change: that is, the activist networks have undertaken a range of political strategies that facilitate change both inside government and in society to advance carbon neutrality. Moreover, it also demonstrates that there is a wide range of expertise located within these carbon-neutrality networks that can directly inform government policy. However, for this to succeed, government would need to be open to changing with whom and how they develop carbon-neutrality policies.

The Cities for Climate Protection Program (CCP) initiative began in 1993, when municipal leaders met (physically) at the United Nations headquarters in New York and adopted a declaration that 'called for the establishment of a worldwide movement of local governments' to 'reduce greenhouse gas emissions, improve air quality, and enhance urban sustainability'.[23] It is a prime example of the ways in which institutional actors have coalesced across distance and location to address a compelling global 'problem'. Since 1993, more than 650 local governments have become involved in the campaign, and this includes 19 councils in Aotearoa/New Zealand, including Wellington and Auckland.[24] The programme is an attempt to provide city and regional local governments with a framework of policy directives and it relies on participants to share information in the virtual world. The ambition of the CCP is to 'build and support a world-wide movement of local governments to achieve tangible improvements in global environmental and sustainable development conditions through cumulative local actions'.[25] It offers another example of how new technologies and activist networks can inform not only radical protest strategies but also local initiatives that are linked transnationally. In other words, while virtually-connected networks can enable wide-ranging, transnational protest action, they can also result in policy changes at all levels of government.

What is evident, then, is that the transfer of ideas and tactics can occur in several ways: between countries, as in the spread of the idea of carbon-neutral cities; and between groups within a movement, such as the Camps for Climate Change or Melbourne's Sustainable Living Festival.[26] We might not always see colourful banners pronouncing social change, but this does not indicate the absence of transformative action around carbon neutrality.[27] We have little historical precedence of less visible, free-roaming networks that undertake their 'work away from the [traditional] halls of power' (such as parliament).[28]

Moreover, the effectiveness of activism can also be measured by the uptake of a network's ideas and values in the wider society, success being indicated by them having permeated the popular culture. The proliferation and penetration of ideas of carbon neutrality, once perhaps regarded as an extreme vision, is demonstrated locally and transnationally in a number of ways. We can see, for example, the influence of Al Gore's movie *An Inconvenient Truth*,[29] which has been seen by millions worldwide, as were the recent Live Earth concerts. Hollywood's A-rated stars are now being ranked according to their environmental commitment. Closer to home, the push for local actions to advance carbon neutrality has been taken up by the TV3 television programme, *Wa$ted*, which is described as:

> ... the show that transforms your average eco crim into your ultimate green convert by taking New Zealand families, auditing their waste and energy usage with the show's unique eco calculator and confronting them with the terrifying truth about their long-term impact on the planet.[30]

In other words, the work of local and transnational networks in sharing ideas and information can transform our consciousness about the importance of carbon neutrality as both an environmental and a political issue.

MAXIMISING OUR DEMOCRATIC RIGHTS

It was mentioned earlier that politicians of all persuasions are keen to be seen as aware of the various environmental issues that are of concern to voters. This indicates that, alongside local and global strategies, focusing our political activism on our national parliament can also make a difference. There are a number of well-known strategies that many of us already employ, including lobbying our local MPs, writing letters to the editor, making submissions and so on.[31] Indeed, we could combine these strategies and persuade our local MPs to draft a series of Members' bills on carbon neutrality. Because there are always more Members' bills proposed than there is time to consider them, a ballot system is used to choose the bills that are introduced. The ballot may contain around 40 drafted bills, but only four will be drawn at any one time. Hence, the need for us to ensure that several bills concerning carbon neutrality are included in the ballot. In a sense, it is like buying a whole book of raffle tickets to increase the chances of winning a prize. The beauty of this strategy is that if a bill is brought before parliament, it must then go to a select committee for public scrutiny. This gives us the opportunity to engage in a wide-ranging public debate and in doing so we can draw from our information-rich carbon-neutral networks in writing up critical submissions, and presenting these to parliament. While few Members' bills become laws, they may affect the government's law-making priorities if they attract sufficient support, so it is a strategy worthy of consideration.[32]

There is one final democratic mechanism available to us all that is often forgotten and that is unique to Aotearoa/New Zealand. In 1993, the Citizens' Initiated Referenda Act was passed into law. Citizens' Initiated Referenda are a form of direct democracy and the exciting aspect of them is that we can use them to focus on a single issue – something that is rarely possible in elections; and it is a channel of influence ideally suited to making use of the wide range of environmental networks already in existence.

The Citizens' Initiated Referenda Act allows for any one of us to trigger a vote on any issue that concerns us. In order to do so, we must submit a proposed question to the Clerk of the House of Representatives. After the question is approved, we have 12 months to collect the signatures of 10 per cent of enrolled electors, which at present equates to 285,189 signatures.[33] If sufficient (valid) signatures are collected, a referendum must be held within 12 months, with public debate on the issue occurring in the lead-up to the vote.[34]

Citizens' Initiated Referenda are rare in liberal democracies. At the national level, only Italy, Switzerland and Aotearoa/New Zealand allow their citizens to initiate a vote (it is used in 24 states in the USA). However, only in Aotearoa/New Zealand are voters allowed to initiate a referendum on any issue of concern: the Swiss may use it only for constitutional matters and in Italy to initiate a change in existing law. Aotearoa/New Zealand is also unique in that the result of the referendum is non-binding on governments. While this may mean that governments could ignore the outcome of a referendum, this is unlikely if the outcome is a clear demonstration of our collective preference. Thus the Citizens' Initiated Referenda mechanism could offer us a political means to force the issue of carbon neutrality onto the policy agenda – demanding that government view it as a pressing national concern requiring immediate and meaningful policy action.

So, why has this not happened already? Certainly the collection of over a quarter of a million signatures is no easy task, and most referendum proposals have lapsed because of an inability to gather the required number of signatures by the due date. Some scholars have suggested that the 10 per cent quota ensures that only well-resourced interest groups are likely to succeed. Norm Withers was successful as an individual in initiating a referendum on justice reform, but realists (or perhaps pessimists) would say he was the exception rather than the rule, because he dedicated himself full time to the gathering of signatures.[35]

There have been 35 referendum proposals submitted to the Clerk of the House since the passage of the act in 1993; 19 of these were submitted between 1993 and 1997, three of which addressed environmental issues.[36] But there have been no proposals concerning the environment submitted in the past 10 years, despite increasing public awareness and concern around climate change and carbon neutrality over the same period. And yet, in the late 1960s, the campaign to save Lake Manapouri culminated in a petition presented to parliament in 1970, signed by 264,907 people, the largest petition in Aotearoa/New Zealand history to that date. Christine Dann reminds us that the issue made a significant contribution to the defeat of the National government at the 1972 election, because the incoming Labour government had pledged to save the lake.[37] This suggests that we New Zealanders can be mobilised in sufficient numbers to add our signatures to a petition demanding a

CIR on carbon neutrality. Given the existence of a multitude of activist networks committed to carbon neutrality, most of which have virtual connections with their membership and the broader public, harnessing a quarter of a million supporters' signatures could be a real possibility.

CONCLUSIONS

Aotearoa/New Zealand has a long history of strong political activism around environmental and human rights concerns. Our reputation for fighting against nuclear-testing in the Pacific and our involvement in the anti-apartheid campaign gained international recognition and support. It may well be that Helen Clark's vision of Aotearoa/New Zealand as the first carbon-neutral nation is realised, but this will not happen without political pressure from outside the 'halls of power'. As we have seen in other parts of this book, carbon neutrality is an urgent issue, one that we can and must act on as part of a broader effort to curb climate change. It is this urgency that compels us here in Aotearoa/New Zealand to take individual and collective actions to be carbon neutral by 2020.

We can effect political change through a wide variety of means, some of which have been demonstrated in this chapter. We can, for example, come together to develop political tactics and ways of making changes in our households, workplaces, schools and communities towards carbon neutrality, drawing on ideas from political networks here but also transnational ones. We can also raise awareness within our communities and take issues to our local and national politicians. We can use our voting power. And, we can initiate a referendum demanding policies for carbon neutrality.

Here it is suggested that the existence of multiple and diverse networks, both locally and transnationally, offers us as New Zealanders a range of possibilities for enhanced and consolidated action around carbon neutrality that can affect, if not revolutionise, the local, national and global campaigns for policy change in this area. These strategies have one thing in common: they are all forms of direct action. They vary in their degree of radicalism, but they are sufficiently diverse modes of activism to appeal to us all as citizens, irrespective of where we sit on the political spectrum or for whom we might vote. Most importantly, successful policy change requires all of us to make a commitment to sustained political participation if we are to realise our goal of carbon neutrality by 2020.

ENDNOTES
1. Downs, A. 1972. Up and down with ecology – 'the issue attention cycle'. *The Public Interest 28*, 38–50.
2. Dann, C. 2001. The Environmental Movement. In Miller, R. (Ed.). *New Zealand Government and Politics*. (pp. 342–351), Auckland: Oxford University Press.
3. Dann, C. 2001.
4. Bale, T. and Wilson, J. 2006. The Greens. In Miller, R. (Ed.), *New Zealand Government and Politics*. (pp. 392–404), Auckland: Oxford University Press.
5. The Green Party website: http://www.greens.org.nz/ (accessed 10 July 2007).
6. Clark, H. 2007. The Prime Minister's Statement to Parliament. 13 February 2007. http://www.beehive.govt.nz/ViewDocument.aspx?DocumentID=28357 (accessed 10 July 2007).
7. See http://www.mfe.govt.nz/issues/climate/take-action/index.html and http://www.4million.org.nz/climatechange/stories/index.php (accessed 10 July 2007).
8. Smith, N. 2006. A BlueGreen Vision for New Zealand. Speech to the National Party Lower North Island Regional Conference, Palmerston North. 13 May 2006. http://www.national.org.nz/Article.aspx?ArticleID=6473. (accessed 10 July 2007).
9. See, for example, Doherty, B. and Doyle, T. 2006. Beyond borders: Transnational politics, social movements and modern environmentalisms. *Environmental Politics 15(5)*, 697–712.
10. Hinchliffe, S. 1997. Home-made space and the will to disconnect. In Hetherington, K. Munro, R. (Eds.). *Ideas of Difference: Social Spaces and the Labour of Division*. Oxford: Blackwell Publishers/The Sociological Review; see also Escobar, A. 1999. Gender, place and networks: a political ecology of cyberculture. In Harcourt, W. (Ed.). *Women@Internet: Creating Cultures in Cyberspace*. London: Zed.
11. See, for example, GreenPages. 2007. *Directory of Environmental and Conservation Groups in Aotearoa/New Zealand* http://www.greenpages.org.nz/ (accessed 02 Aug 2007).
12. anon., cited in Byrne, P. 1997. *Social Movements in Britain*. London: Routledge, p. 146.
13. Mountier, F. 2005. Save Happy Valley Coalition press release: http://www.savehappyvalley.org.nz/pr_22-12-05_shvc (accessed 24 Aug 2007).
14. Keck, M. and Sikkink, K. 1998. Transnational advocacy networks in the movement society. In Meyer, D.S. Tarrow, S. (Eds). *The Social Movement Society: Contentious Politics for a New Century*. Lanham, ML: Rowman and Littlefield.
15. Wellman, B. and Gulia, M. 1999. Virtual communities as communities: net surfers don't ride alone. In Smith, M.A. and Kollock P. (Eds) *Communities in Cyberspace*. London: Routledge, p. 187.
16. Doherty, B. and Doyle, T. 2006.
17. For example, FOE.2006. *FOE Welcome New Climate Change Information Packs*. Activist Network posting: http://www.activistnetwork.org.uk/pn/modules.php?op=modload&name=News&file=article&sid=747 (accessed 07 Aug 2007).
18. Network for Climate Change 2006. http://www.climatecamp.org.uk/ (accessed 07 Aug 2007).
19. Network for Climate Change 2007; see also Taylor, J. 2007. 'Climate campaigners glue themselves to transport building.' *The New Zealand Herald*, 17 August. http://www.nzherald.co.nz/feature/story.cfm?c_id=26&objectid=10458530 (accessed 20 Aug 2007).
20. EarthFirst! Netherlands. 2007. Shrink or Drown Dutch/Belgian EarthFirst! Gathering 2007. Information available at http://www.climatecamp.org.uk/dutch_flyer_en.gif. (accessed 02 Aug 2007); Southeast Convergence for Climate Action. 2007. website: http://www.climateconvergence.org/southeast/ (accessed 02 Aug 2007); West Coast Convergence for Climate Action. 2007. website: http://www.climateconvergence.org/west. (accessed 02 Aug 2007).
21. Friends of the Earth. 2006.

22 Friends of the Earth, 2006.
23 ICLEI. 2007. Cities for Climate Change Program: ICLEI Global Program. http://www.iclei.org/documents/Global/brochures/ICLEI_Brochuretext_ENG.pdf (accessed 10 Aug 2007), pp. 1–2.
24 ICLEI Oceania. 2007. Communities for Climate Protection New Zealand http://www.iclei.org/index.php?id=1387&region=OC; (accessed 10 Aug 2007); Kedgley, S. 2007. Green Party Wellington Transport Spokesperson Press Release. Available at http://www.greens.org.nz/searchdocs/PR10918.html (accessed 02 Aug 2007).
25 Europe Cities for Climate Protection Campaign. 2007. For more information, visit the website: http://www.managenergy.net/actors/A1420.htm. (accessed 28 July 2007); Slocum, R. 2004. Polar bears and energy-efficient lightbulbs: strategies to bring climate change home. *Environment and Planning D: Society AND Space 4*; cf. Betsill, M.M. and Bulkeley, H. 2004. Transnational networks and global environmental governance: the cities for climate protection program. *International Studies Quarterly 48*, 471–493.
26 Information for these events is available at the following websites: Camp for Climate Action. 2007. http://www.climatecamp.org.uk/ (accessed 27 July 2007); Sustainable Living Festival. 2007. http://www.slf.org.au/festival/ (accessed 11 Apr 2007).
27 Castells, M. 2000. *The Rise of the Network Society: The Information Age – Economy, Society and Culture Volume 1* (2nd edn). Malden, MA: Blackwell.
28 Castells, M. 2000: 362.
29 *An Inconvenient Truth* (motion picture). 2006. Paramount Classics and Participant Productions.
30 TV3. 2007. TV3 Programme Listing website: http://www.tv3.co.nz/Programmes/Wasted/ (accessed 10 Aug 2007).
31 To find out more, see Hughes, F. and Calder, S. 2007. *Have Your Say. Influencing public policy in New Zealand*. Wellington: Dunmore Publishing; Young, A. 2003. *The Good Lobbyist's Guide*. Auckland: Exisle Publishing.
32 Laila Harre's private members bill on Paid Parental Leave was critical to the adoption and eventual implementation of a paid parental leave policy by the Labour-Alliance Coalition Government in 2002.
33 As at 20 August 2007, there were 2,851,896 voters on the electoral roll.
34 Catt, H. 2001. Citizens' Initiated Referenda. In Miller, R. (Ed.). *New Zealand Government and Politics*. (pp. 3386–3407), Auckland: Oxford University Press. Parkinson, J. 2006. Direct Democracy. In Miller, R. (Ed.). *New Zealand Government and Politics*. (pp. 547–561), Auckland: Oxford University Press.
35 Catt, H.2001.
36 Clerk of the House of Representatives, unpublished data, 2007.
37 Dann, C. 2001.

Conclusion

This book shows how New Zealanders can take on the carbon-neutral challenge. We urge anyone who recognises the importance of this issue to act now. At the same time, we urge the government to immediately introduce legislation, policies and incentives that will show that this country has the foresight and courage to lead the world on what may well be the most important issue of our time.

Many people are cynical about democracy and believe that power is held by a few. We disagree with this view. Power lies in the actions of ordinary people, everywhere. Together, we keep the system going, and together we can change it.

Below are suggestions for action.

INDIVIDUAL ACTIONS
1. **Transport** *Stay local.* The activity that generates the fewest carbon emissions is walking, followed by cycling. Going to school, working and shopping near home reduces the need to use a motor vehicle to get around. When people walk and cycle, vibrant communities are created and traffic congestion and air pollution are reduced. If you need to go across town, buses and trains are usually better options than cars, and if you need to take a car consider sharing your trip with others.
2. **Spending** *Think of spending as voting.* Every time you part with your cash, you are casting a vote in favour of what you just bought. First, think about whether you really need the product you are reaching for, especially if it is made from limited resources (such as plastic goods, technical gadgets, synthetic clothing) and/or involves carbon-intensive manufacture and distribution (like drinks in aluminium cans, air-freighted imported goods). Second, buy food and goods you think are viable in a carbon-neutral, sustainable world (for example, organic food, fair-trade gifts).

3. **Investing** *Put your money where your values are.* Check what is happening to money being invested on your behalf – in a bank account, in an investment fund, in a KiwiSaver scheme. Ask your provider what the organisation's ethical and responsible investment policies are. If these are not to your liking, ask that they be changed.
4. **Your home** *Stay put.* Live in one house for a long time. Resist renovating on the basis of fashion. Do not indulge in do-ups or short-term fixes that will result in higher levels of carbon and other toxins being released into the environment. Invest in improvements that will save carbon emissions, such as keeping your home warm using less energy or creating a home office so you do not have to travel.
5. **Your garden** *Have one.* Compost your food scraps. Grow organic food and share it with your neighbours. Do not get disheartened if the snails get most of your vegetables at first.
6. **Your children's education** *Support your local schools.* Ask what their policies are on resource conservation and how environmental action is integrated into the curriculum. Or become a school board member and create the policies.

ORGANISATIONAL ACTIONS
1. In most organisations, such as workplaces, schools, sports clubs and community groups, there is a space where people can have their say – take that opportunity.
2. Strive for policy changes that will ensure the organisation will take into consideration the environmental impact of all decisions.
3. Be on the look out for opportunities to initiate low-carbon practices and say that that is what you are doing. For example, suggest meetings be held by teleconferencing to save the carbon generated by car and air travel. Put up notices reminding people to turn out lights.
4. If you are a leader, consider whether the carbon-neutral challenge could be a unifying, positive project for your organisation. Try things out, even if you are uncertain that they are the best action to take. Check out carbon offsetting programmes, like CarboNZero. These are great ways to measure your emissions as well as to put money into carbon related projects.
5. Develop a partnership with a local environmental organisation. There are many grass roots groups making a difference in their communities and, by teaming up, both organisations can gain new knowledge and create positive action.

POLITICAL ACTIONS
1. Write to newspapers and magazines, your local MP and the leaders of political parties asking for legislation, policies and programmes you believe will make a difference.
2. Ensure you are enrolled to vote so you can vote in local government and national elections as well as referenda (http://www.elections.org.nz/voter-enrolment.htmls). Then make sure you cast your vote, and vote for people and parties that have clear climate change policies you think will work.
3. Ask candidates in national and local body elections about their climate change views and policies, and inform them that this issue is a high priority for you and your vote.
4. Make submissions to your local authority's annual plan suggesting climate change practices that can be integrated with council projects.
5. Turn up to climate change events. This helps strengthen the organisations that run these events, and sends the message to politicians that people want action.
6. Consider joining a political party. If you are happy with the party's climate change policies, help it get elected. If you are unhappy with its policies, try and change them from within.
7. Join an organisation that campaigns for climate change action. Give it money or time or both.
8. Investigate if there are on-line or Internet-based networks, local or global, that you can participate in 'virtually' in order to protest against carbon emissions.

Let's get serious about making Aotearoa/New Zealand the first carbon-neutral nation leading the world into a sustainable future.

APPENDICES

Appendix 1

Note: These figures are reprinted exactly as they appeared in their original source: Denne, T. 2006. "Achieving emissions reductions at least cost: international lessons for the New Zealand context." Presentation at *Climate Change: The Policy Challenges Symposium*, Victoria University Institute of Policy Studies. 6 Oct 2006. Slides available at http://ips.ac.nz/events/completed-activities/Climate%20Change%20Symposium/Proceedings.html

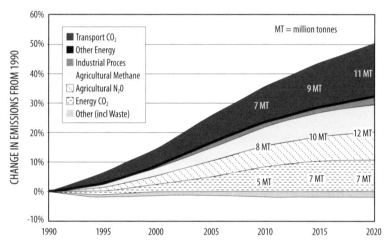

Fig. 1 Observed and predicted growth in greenhouse gas emissions in New Zealand since 1990

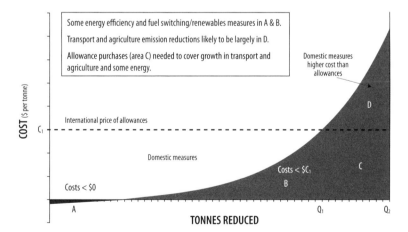

Fig. 2 Costs to New Zealand of emissions reductions

Appendix 2

THE EARTH CHARTER

PREAMBLE

We stand at a critical moment in Earth's history, a time when humanity must choose its future. As the world becomes increasingly interdependent and fragile, the future at once holds great peril and great promise. To move forward we must recognize that in the midst of a magnificent diversity of cultures and life forms we are one human family and one Earth community with a common destiny. We must join together to bring forth a sustainable global society founded on respect for nature, universal human rights, economic justice, and a culture of peace. Towards this end, it is imperative that we, the peoples of Earth, declare our responsibility to one another, to the greater community of life, and to future generations.

Earth, Our Home

Humanity is part of a vast evolving universe. Earth, our home, is alive with a unique community of life. The forces of nature make existence a demanding and uncertain adventure, but Earth has provided the conditions essential to life's evolution. The resilience of the community of life and the well-being of humanity depend upon preserving a healthy biosphere with all its ecological systems, a rich variety of plants and animals, fertile soils, pure waters, and clean air. The global environment with its finite resources is a common concern of all peoples. The protection of Earth's vitality, diversity, and beauty is a sacred trust.

The Global Situation

The dominant patterns of production and consumption are causing environmental devastation, the depletion of resources, and a massive extinction of species. Communities are being undermined. The benefits of development are not shared equitably and the gap between rich and poor is widening. Injustice, poverty, ignorance, and violent conflict are widespread and the cause of great suffering. An unprecedented rise in human population has overburdened ecological and social systems. The foundations of global security are threatened. These trends are perilous—but not inevitable.

The Challenges Ahead

The choice is ours: form a global partnership to care for Earth and one another or risk the destruction of ourselves and the diversity of life. Fundamental changes are needed in our values, institutions, and ways of living. We must realize that when basic needs have been met, human development is primarily about being more, not having more. We have the knowledge and technology to provide for all and to reduce our impacts on the environment. The emergence of a global civil society is creating new opportunities to build a democratic and humane world. Our environmental, economic, political, social, and spiritual challenges are interconnected, and together we can forge inclusive solutions.

Universal Responsibility

To realize these aspirations, we must decide to live with a sense of universal responsibility, identifying ourselves with the whole Earth community as well as our local communities. We are at once citizens of different nations and of one world in which the local and global are linked. Everyone shares responsibility for the present and future well-being of the human family and the larger living world. The spirit of human solidarity and kinship with all life is strengthened when we live with reverence for the mystery of being, gratitude for the gift of life, and humility regarding the human place in nature.

We urgently need a shared vision of basic values to provide an ethical foundation for the emerging world community. Therefore, together in hope we affirm the following interdependent principles for a sustainable way of life as a common standard by which the conduct of all individuals, organizations, businesses, governments, and transnational institutions is to be guided and assessed.

PRINCIPLES

I. RESPECT AND CARE FOR THE COMMUNITY OF LIFE

1. Respect Earth and life in all its diversity.
a. Recognize that all beings are interdependent and every form of life has value regardless of its worth to human beings.
b. Affirm faith in the inherent dignity of all human beings and in the intellectual, artistic, ethical, and spiritual potential of humanity.

2. Care for the community of life with understanding, compassion, and love.
a. Accept that with the right to own, manage, and use natural resources comes the duty to prevent environmental harm and to protect the rights of people.
b. Affirm that with increased freedom, knowledge, and power comes increased responsibility to promote the common good.

3. Build democratic societies that are just, participatory, sustainable, and peaceful.
a. Ensure that communities at all levels guarantee human rights and fundamental freedoms and provide everyone an opportunity to realize his or her full potential.
b. Promote social and economic justice, enabling all to achieve a secure and meaningful livelihood that is ecologically responsible.

4. Secure Earth's bounty and beauty for present and future generations.
a. Recognize that the freedom of action of each generation is qualified by the needs of future generations.
b. Transmit to future generations values, traditions, and institutions that support the long-term flourishing of Earth's human and ecological communities. In order to fulfill these four broad commitments, it is necessary to:

II. ECOLOGICAL INTEGRITY

5. Protect and restore the integrity of Earth's ecological systems, with special concern for biological diversity and the natural processes that sustain life.
a. Adopt at all levels sustainable development plans and regulations that make environmental conservation and rehabilitation integral to all development initiatives.
b. Establish and safeguard viable nature and biosphere reserves, including wild lands and marine areas, to protect Earth's life support systems, maintain biodiversity, and preserve our natural heritage.
c. Promote the recovery of endangered species and ecosystems.
d. Control and eradicate non-native or genetically modified organisms harmful to native species and the environment, and prevent introduction of such harmful organisms.
e. Manage the use of renewable resources such as water, soil, forest products, and marine life in ways that do not exceed rates of regeneration and that protect the health of ecosystems.
f. Manage the extraction and use of non-renewable resources such as minerals and fossil fuels in ways that minimize depletion and cause no serious environmental damage.

6. Prevent harm as the best method of environmental protection and, when knowledge is limited, apply a precautionary approach.
a. Take action to avoid the possibility of serious or irreversible environmental harm even when scientific knowledge is incomplete or inconclusive.
b. Place the burden of proof on those who argue that a proposed activity will not cause significant harm, and make the responsible parties liable for environmental harm.
c. Ensure that decision making addresses the cumulative, long-term, indirect, long distance, and global consequences of human activities.
d. Prevent pollution of any part of the environment and allow no build-up of radioactive, toxic, or other hazardous substances.
e. Avoid military activities damaging to the environment.

7. Adopt patterns of production, consumption, and reproduction that safeguard Earth's regenerative capacities, human rights, and community well-being.
a. Reduce, reuse, and recycle the materials used in production and consumption systems, and ensure that residual waste can be assimilated by ecological systems.
b. Act with restraint and efficiency when using energy, and rely increasingly on renewable energy sources such as solar and wind.
c. Promote the development, adoption, and equitable transfer of environmentally sound technologies.

d. Internalize the full environmental and social costs of goods and services in the selling price, and enable consumers to identify products that meet the highest social and environmental standards.
e. Ensure universal access to health care that fosters reproductive health and responsible reproduction.
f. Adopt lifestyles that emphasize the quality of life and material sufficiency in a finite world.

8. Advance the study of ecological sustainability and promote the open exchange and wide application of the knowledge acquired.
a. Support international scientific and technical cooperation on sustainability, with special attention to the needs of developing nations.
b. Recognize and preserve the traditional knowledge and spiritual wisdom in all cultures that contribute to environmental protection and human well-being.
c. Ensure that information of vital importance to human health and environmental protection, including genetic information, remains available in the public domain.

III. SOCIAL AND ECONOMIC JUSTICE

9. Eradicate poverty as an ethical, social, and environmental imperative.
a. Guarantee the right to potable water, clean air, food security, uncontaminated soil, shelter, and safe sanitation, allocating the national and international resources required.
b. Empower every human being with the education and resources to secure a sustainable livelihood, and provide social security and safety nets for those who are unable to support themselves.
c. Recognize the ignored, protect the vulnerable, serve those who suffer, and enable them to develop their capacities and to pursue their aspirations.

10. Ensure that economic activities and institutions at all levels promote human development in an equitable and sustainable manner.
a. Promote the equitable distribution of wealth within nations and among nations.
b. Enhance the intellectual, financial, technical, and social resources of developing nations, and relieve them of onerous international debt.
c. Ensure that all trade supports sustainable resource use, environmental protection, and progressive labor standards.
d. Require multinational corporations and international financial organizations to act transparently in the public good, and hold them accountable for the consequences of their activities.

11. Affirm gender equality and equity as prerequisites to sustainable development and ensure universal access to education, health care, and economic opportunity.
a. Secure the human rights of women and girls and end all violence against them.
b. Promote the active participation of women in all aspects of economic, political, civil, social, and cultural life as full and equal partners, decision makers, leaders, and beneficiaries.
c. Strengthen families and ensure the safety and loving nurture of all family members.

12. Uphold the right of all, without discrimination, to a natural and social environment supportive of human dignity, bodily health, and spiritual well-being, with special attention to the rights of indigenous peoples and minorities.
a. Eliminate discrimination in all its forms, such as that based on race, color, sex, sexual orientation, religion, language, and national, ethnic or social origin.
b. Affirm the right of indigenous peoples to their spirituality, knowledge, lands and resources and to their related practice of sustainable livelihoods.
c. Honor and support the young people of our communities, enabling them to fulfill their essential role in creating sustainable societies.
d. Protect and restore outstanding places of cultural and spiritual significance.

IV. DEMOCRACY, NONVIOLENCE, AND PEACE

13. Strengthen democratic institutions at all levels, and provide transparency and accountability in governance, inclusive participation in decision making, and access to justice.
a. Uphold the right of everyone to receive clear and timely information on environmental matters and all development plans and activities which are likely to affect them or in which they have an interest.
b. Support local, regional and global civil society, and promote the meaningful participation of all interested individuals and organizations in decision making.
c. Protect the rights to freedom of opinion, expression, peaceful assembly, association, and dissent.
d. Institute effective and efficient access to administrative and independent judicial procedures, including remedies

and redress for environmental harm and the threat of such harm.
e. Eliminate corruption in all public and private institutions.
f. Strengthen local communities, enabling them to care for their environments, and assign environmental responsibilities to the levels of government where they can be carried out most effectively.

14. Integrate into formal education and life-long learning the knowledge, values, and skills needed for a sustainable way of life.
a. Provide all, especially children and youth, with educational opportunities that empower them to contribute actively to sustainable development.
b. Promote the contribution of the arts and humanities as well as the sciences in sustainability education.
c. Enhance the role of the mass media in raising awareness of ecological and social challenges.
d. Recognize the importance of moral and spiritual education for sustainable living.

15. Treat all living beings with respect and consideration.
a. Prevent cruelty to animals kept in human societies and protect them from suffering.
b. Protect wild animals from methods of hunting, trapping, and fishing that cause extreme, prolonged, or avoidable suffering.
c. Avoid or eliminate to the full extent possible the taking or destruction of non-targeted species.

16. Promote a culture of tolerance, nonviolence, and peace.
a. Encourage and support mutual understanding, solidarity, and cooperation among all peoples and within and among nations.
b. Implement comprehensive strategies to prevent violent conflict and use collaborative problem solving to manage and resolve environmental conflicts and other disputes.
c. Demilitarize national security systems to the level of a non-provocative defense posture, and convert military resources to peaceful purposes, including ecological restoration.
d. Eliminate nuclear, biological, and toxic weapons and other weapons of mass destruction.
e. Ensure that the use of orbital and outer space supports environmental protection and peace.
f. Recognize that peace is the wholeness created by right relationships with oneself, other persons, other cultures, other life, Earth, and the larger whole of which all are a part.

THE WAY FORWARD

As never before in history, common destiny beckons us to seek a new beginning. Such renewal is the promise of these Earth Charter principles. To fulfill this promise, we must commit ourselves to adopt and promote the values and objectives of the Charter.

This requires a change of mind and heart. It requires a new sense of global interdependence and universal responsibility. We must imaginatively develop and apply the vision of a sustainable way of life locally, nationally, regionally, and globally. Our cultural diversity is a precious heritage and different cultures will find their own distinctive ways to realize the vision. We must deepen and expand the global dialogue that generated the Earth Charter, for we have much to learn from the ongoing collaborative search for truth and wisdom.

Life often involves tensions between important values. This can mean difficult choices. However, we must find ways to harmonize diversity with unity, the exercise of freedom with the common good, short-term objectives with long-term goals. Every individual, family, organization, and community has a vital role to play. The arts, sciences, religions, educational institutions, media, businesses, nongovernmental organizations, and governments are all called to offer creative leadership. The partnership of government, civil society, and business is essential for effective governance.

In order to build a sustainable global community, the nations of the world must renew their commitment to the United Nations, fulfill their obligations under existing international agreements, and support the implementation of Earth Charter principles with an international legally binding instrument on environment and development.

Let ours be a time remembered for the awakening of a new reverence for life, the firm resolve to achieve sustainability, the quickening of the struggle for justice and peace, and the joyful celebration of life.

ORIGIN OF THE EARTH CHARTER
The Earth Charter was created by the independent Earth Charter Commission, which was convened as a follow-up to the 1992 Earth Summit in order to produce a global consensus statement of values and principles for a sustainable future. The document was developed over nearly a decade through an extensive process of international consultation, to which over five thousand people contributed. The Charter has been formally endorsed by thousands of organizations, including UNESCO and the IUCN (World Conservation Union). For more information, please visit www.EarthCharter.org.

Appendix 3

PUBLISHER'S NOTE

As part of our commitment to publishing this book, we were keen to do whatever we reasonably could to minimise the CO_2 emissions involved in its production and distribution. Apart from the obvious matter of putting our money where our mouth is, I also felt that it would be a good first step in initiating the much bigger process of taking our company toward carbon neutrality.

Grappling with our carbon footprint has, predictably, proved to be complex, at times daunting, but also inspiring. It had a good start – when I first talked at a staff meeting about this book and the issue of CO_2 emissions, I was hugely impressed by the immediate and passionate support from my co-workers. The palpable goodwill and enthusiasm for tackling this issue is hugely encouraging, as this kind of buy-in makes long-lasting change far more likely.

The obvious starting point in reducing the CO_2 emissions involved in the production of a book is an assessment of the impact of the office environment from where it emanates. In terms of the whole project, that impact is probably relatively limited – our offices are modest in scale and fittings, our energy use is reasonable, we recycle wherever we can, our bike rack is well used, a number of staff regularly walk to work – though we have yet to do a detailed audit on our footprint, so we can claim little credit in this area.

With the book's editors based in Auckland and England, and our offices in Nelson, travel could potentially have posed an issue. But aside from one meeting with the Auckland-based editor (which piggy-backed on four other meetings), book production has been conducted entirely via email and the phone.

The physical manufacturing of the book posed some of the most obvious challenges. That said, one pivotal part of the process that has been enormously satisfying and surprisingly easy has been finding appropriate paper stock. The paper between your fingers comes from the Advance Agro Mill in Thailand, the first integrated paper mill in South East Asia to receive ISO14001 accreditation. The paper is made from sustainably farmed, quick-growing eucalyptus trees, and is carbon neutral, as the trees naturally offset the emissions from the paper mill. Transporting the paper to New Zealand is obviously a significant factor, though we mitigated this slightly by sourcing paper from Asia, rather from Europe, where carbon-neutral paper of a higher quality can be sourced.

Carbon Neutral by 2020 was printed in Wellington, also to minimise transport distances, by Astra Print Ltd. They are currently working through the certification levels of Enviro-Mark and have made a number of changes which are very relevant to this process. These include; using mineral oil-free inks, based on 100% renewable resources; use of CFC-free press chemicals; printing on machinery that is mostly the latest generation from Heidelberg, economical to run and complying to the highest European environmental standards, while also developed to produce the minimum of waste; reducing waste collection from daily pickups to three times a week; on site recycling bins for waste plastic and shrinkwrap, as well as most waste paper and offcuts.

Perhaps the most difficult area in which to make any progress in the publishing process has proved to be the physical method of distribution – getting the books from our warehouses to the bookshops that we supply all over New Zealand. Here the pressure of commercial reality really does rear its head. An essential part of being a successful publisher is having good customer service, and delivering books quickly and reliably in-store is the foundation of that. We use couriers for these deliveries, with all of the obvious CO_2 emissions associated with the vans, truck and planes that underpin this service. Right now it is not obvious how we could avoid this, and stay in business.

I had always imagined, naively as it happens, that as least as far as *Carbon Neutral by 2020* is concerned, I would be able to offset transport involved by purchasing carbon credits. To my surprise, this is not straightforward. Landcare's CarboNZero creditation programme, which we are currently investigating for our whole publishing operation, only offers the ability for businesses to offset their emissions once they are part of the programme, which as yet we are not.

They do however, have a Travel and Tourism Calculator which allows travellers to calculate and offset their CO_2 emissions. While not designed for business offsetting, I am nonetheless using this calculator to make some rough estimates of the travel involved in the production and distribution of this book, and will buy the requisite carbon credits. Clearly, we still have work to do.

Robbie Burton
Managing Director and Publisher

Index

Aarhus 122
Accident Compensation Corporation 126
action-based learning 34–6
activism. *See* political activism
activist investors 224–5, 226–7
advertising 191
agriculture
    and greenhouse gas emissions 12, 14, 166
    deep organics 171, 172
    efficiency model 171, 172
    energy ratio 90
    industrial model 171, 172
    organic 164–83, 242, 255–6
    shallow organics 171, 172
air travel 13, 80, 81, 95, 99, 147, 148–9, 150, 238
Alesina, Nina 195
alternative investors 225, 226, 227
altruism 17–18, 197. *See also* care
aluminium joinery 48–9
Amazon rainforest 11
AMP, Australia 222–3
Anderson, Ray 230–1
Antarctic, sea ice shrinkage 10
appliances, electric 12, 50–1, 52, 62, 84–5
    Energy Star system 50
    sharing 82, 83
architects 30, 36–7, 43, 68, 70, 72
Arctic, sea ice shrinkage 10
Armstrong, Karen 22
Asia 132, 133, 134–5, 166
Asian population in NZ 82
Asteron Socially Responsible Investment Trust 227, 229
Auckland 83, 93, 105–6, 107, 168, 176, 285
    Permaculture demonstration home 177–8
    reducing carbon burden from transport system 113–30
Auckland City Council 121, 124
Auckland Regional Council 118, 125–6, 271–2
Auckland Regional Growth Strategy 119, 123–4
Auckland Regional Policy Statement 118–19
Auckland Regional Transport Authority 107
Australia 50, 104, 123, 124, 139, 204, 222–3, 261
Australian Stock Exchange 223
automobile dependence 101, 102, 104, 106, 109, 110
avoidance investors 224, 226

Banks, John 93
bathrooms 45, 49, 51–2, 55, 58–9, 71, 73, 85
    rubbish disposal 49, 53
Beachlands Maraetai Resource Depot 169–70
Beacon Pathway Ltd 68, 229

Bhopal disaster 220
biofuels 100, 127–8, 209
bioinformatics 152
Bokashi 169
Bolivia 193
books, electronic 157
Brazil 138, 261
Britain 12, 95, 115, 121–2, 153, 238, 240, 284, 285
Building Act 2004 69, 75
building and property professionals, and creation of carbon-neutral buildings 71–3
Building Information Modelling tools 43–4, 72
building materials 72
    durability 54, 56, 60, 69, 74
    sustainable 56, 69, 74, 75–6
    toxic 54–6, 72, 75–6
Building Research Association of New Zealand (BRANZ) 68, 83
building site waste 69, 72, 74–5, 141
buses 94, 107, 135
    Auckland 123, 124
    carbon dioxide emissions 117, 118
    electric 124–5
business
    irresponsible 219–21
    responsible/sustainable 221–3, 227–8, 235–57
    See also investment, responsible
Business Ethics Awards 231, 232
'bystander effect' 17

cabinetry 55–6
Campaign for Nuclear Disarmament 279
Camps for Climate Action 284–5, 286
Canada 115, 123, 262
car boot sales 87
carbon calculators 250, 254
carbon credits 15, 79–80, 100–1
carbon crisis, New Zealand 9–15
carbon cycle 10
carbon dioxide emissions 9–10, 22, 79
    air travel 13, 80, 95, 99, 147, 238
    atmospheric levels 10, 11
    buildings 41–2, 80
    household 40, 41–2, 58, 59, 64, 71, 79, 80–1, 95
    monitoring 153–4
    New Zealand, statistics 11–13, 295
    transport 13, 92, 93, 94, 98, 99–101, 114, 115, 116–8, 127, 253–4
carbon footprint. *See* carbon dioxide emissions
carbon offsetting 16, 22, 101
carbon sinks 167, 212
    soil 167–8

INDEX

carbon tax, New Zealand 14, 15, 149, 260, 281
carbon-trading scheme 260, 266
carboNZero carbon offsetting scheme 22, 236, 253–4, 293
'care' 18–20
  in context of *Earth Charter* 207
  See also altruism
carpet 47
carpooling 104, 109, 115, 121, 122, 125–6
car-sharing 104–5, 109
cars 10, 79, 81, 93–4, 98, 107–8, 159
  Auckland 114, 115, 125–8, 159
  carbon dioxide emissions 117, 118, 126, 159
  electric 126–7
  energy efficient, hybrid 81, 94–5, 127
  internalising environmental costs 98, 102–8
  ownership 115
  ownership, Auckland 114, 115
  ownership, New Zealand 13, 99
  size 79, 93–4, 126, 127, 128
  See also parking
Catalyst carbon calculator 250, 254
cell phones 146, 161, 191
cheminformatics 152
Chencha people, house design 190
children 24, 38, 82, 83, 92, 93, 94, 98, 132, 146, 180
  involvement in gardening 165, 169, 173, 176
  See also Enviroschools Programme; schools
chimneys 45
China 11, 122, 123, 134–5, 166, 170, 192, 261
Christchurch 178–9
Cities for Climate Protection Program (CPP) 277, 285–6
Citizens' Initiated Referenda Act 287–8
citizenship, sense of 25
Clark, Helen 8, 42, 289
Clean Energy Guide 241
climate change 10
  government policy 8, 14, 15, 25, 151, 242–3, 259, 260, 263, 266, 277, 278, 281
  impact on business 238
  impact on New Zealand 14, 17
  An Inconvenient Truth 19, 124, 200–1, 213–14, 231, 238–9, 286
  international obligations 261–2, 270
  models 9, 10, 11, 16, 151–2, 187
  motivators for taking action 204–5
  personal obstacles to taking action 16–21
  Stern Review of the Economics of Climate Change 101, 187, 241–2
  tipping point 186–7, 236–43
  See also global warming; greenhouse gases
Climate Change Response Act 2002 261
climate neutrality 230–1

clothes driers 53–4
coal 10, 12, 134, 283
coastal areas, impact of climate changes 14
co-housing schemes 83
commuting distance 92, 93
composting 52, 87, 88, 89, 168–70, 176, 177, 178
  cold 169
  hot 169, 177, 178
composting toilets 139–40, 170
computational steering 155
Computer Aided Design (CAD) 152
computer models 10, 11, 16, 146–7
  of climate 9, 10, 11, 16, 151–2, 187
computer simulations 146, 152, 155
computers 62, 145–63
  energy-intensive construction 161
condensation 85
construction materials. *See* building materials
consumers, and creation of carbon-neutral buildings 70–1
Consumer's Institute 68
consumption 12–13, 64, 86–7, 190–3, 194, 195, 237
  ethical spending 211–12, 292
  reducing 79–95, 292
  See also energy consumption; overconsumption; shopping
'contaminant', in Resource Management Act 262–3, 267
continuous improvement cycle 248–52
cooling, in houses 44–5
  passive 46, 62–3
Copenhagen 108
coral reefs, bleaching 10
Covey, Steven, *The Seven Habits of Highly Effective Families* 24
craft 188, 195–7
Creo 68
Crowder, Bob 177, 178–9
Cuba 174
Curitiba 138
cycling 13, 62, 81, 95, 102, 109, 121, 292
  and shopping malls 138
  Auckland 114, 115, 120, 122, 128, 129
  carbon dioxide emissions 95, 117, 118
  E-bikes 158
cyclones 14

Dann, Christine 288
Darlington 121–2
De Leuw Cather and Company 115
debt, level of 132, 133
deforestation. *See* forests, destruction
dehumidifiers 86
Deloitte 231
demolition materials 72
denial 204
Denmark 83, 108, 122

303

Department of Building and Housing 67–8
Department of Conservation 280
design
  as aid to survival 188–90
  as threat to survival 190–3
  sustainable 193–8
  vision for 2020 186
designers 72
diet 90–2. *See also* food
Diet for a Small Planet 91-2
digital photography 156–7
Domine, Andre 172–3
double glazing 46, 47–9, 81
drought, New Zealand 14
Dunlop, Robin 125

Earth Charter 204, 205–10, 296–9
  practical suggestions for using 210–13
  principles 206–7, 208, 212–13, 297
Earth Charter Community Action Tool (EarthCAT) 213
Earth Charter Initiative 212
Earth Charter Youth Initiative 212
Earth Summit 206
EarthFirst! 285
Earthsong Eco-Neighbourhood, Ranui 233
eBay 87
E-bikes 158
E-books 157
Ecology Action 279
economic issues 100, 102–6, 108, 180, 192, 194, 200, 204, 205, 237, 238, 239, 244
  and responsible investing 217, 222, 224, 227, 228, 230, 233
  and the *Earth Charter* 205, 209, 211, 213
  Stern Review of the Economics of Climate Change 101, 187, 241–2
eco-tourism 149–50, 242
education 293
  Earth Charter as educational tool 212
  studying from home 148
  See also Enviroschools programme; schools
El Niño 151
electric vehicles 100
electricians 72
electricity consumption. *See* energy consumption
electricity generation 10, 12, 65–6, 115, 265–6, 269–72
  domestic 63–6, 95, 196, 198
empathy 17
energy. *See also* electricity generation; hydro-electricity; renewable energy resources; solar energy; wind power
energy consumption 12–13, 42, 44, 84–5
  monitoring 153–4
energy efficiency 12, 42, 43, 50–1, 242, 260
  shopping malls (future scenario) 140

Energy Efficiency and Conservation Authority (EECA) 68
Energy Wise Rally 94
Enron 221
Enschede 122
Enviro-Mark® 255
Environment Court 265, 272
'Environmental Choice' scheme 73, 255
Environmental Defence Society v Auckland Regional Council 2002 271–2
environmental footprint 22
environmental management systems 254–5
environmental movement 279–80
Environmental Performance Review of New Zealand (OECD) 239–40
Enviroschools Programme 32–8, 176, 212
  community support 36–7
  living curriculum 34–6
  mission 32
  Principles 33–4
erosion 14, 168, 171
ethics, in climate change strategies 199–215, 259
  necessity for ethical action 201–5
  vision for 2020 200
Europe 11, 83, 91, 95, 123, 153, 221, 238, 240, 262, 284, 285
extinction, threatened 10, 237
Exxon Valdez oil spill 220–1

family, extended 82–3
family identity projects 24–5
'fart tax' 14
feature-tracking technologies 155
ferries, carbon dioxide emissions 117, 118
fertilisers 168, 171, 172, 174
fibreglass batts 47
Filatosa 189
Finite Element Modelling (FEM) 152
fireplaces 45
Fisher & Paykel Healthcare 229
Fitzsimons, Jeanette 260, 271
Flannery, Tim 182
Fletcher Building 229
floods, New Zealand 14, 108, 133
foil insulation 47
food 90–2
  home and locally produced 25, 79, 89, 91, 92, 166, 174, 174–9
  organic 81, 238
  seasonal 90
  storage 52
  See also agriculture; diet; gardening
food industry 90, 91, 165
  dependence on oil 90
  energy ratio 90
  supply chains 91
  See also agriculture
food miles 91, 166, 240

Ford Pinto case 219–20
Forestry Stewardship Council 256
forests, destruction 10, 158, 208, 211–12, 237
    impact on atmospheric greenhouse gases 11, 12, 211–12
    New Zealand 12
    See also Amazon rainforest
formaldehyde 47, 55, 56
fossil fuels 10, 44, 90, 100, 134, 136, 161, 208, 260. *See also* coal; gas; oil
freight transport 99
fridges, energy efficiency 50, 51, 52
Friends of the Earth 280, 285
fruit 90, 91
fuel efficiency standards 104
fuel substitution 100
fuel tax 102, 103, 105

garages 61, 121
gardening 79, 89, 293
    diversity of food crops 175
    organic 169–70, 174–9, 293
    urban gardens 174–9
gas 10, 12
GE Free Food Guide 241
GE-free regions 180
Geldof, Bob 17, 18
Genesis Energy 266
Germany 64, 95, 124
Get Sustainable Challenge 236, 253
glaciers, retreating 10, 16–17, 187
glazing 47–9
    double glazing 46, 47–9, 81
    low emissivity or solar control glass 48
    See also windows
global warming 9, 10–12, 13–14, 16, 237
    impact on New Zealand 14, 17
goods. *See* products
Gore, Al 124, 187, 200–1, 213, 231, 238–9, 286
government policy 194
    building and renovation 73–6
    climate change 8, 14, 15, 25, 151, 242–3, 259, 260, 263, 266, 277, 278, 281
    'contagion' effect 281
    environment 263
    home gardening 176
    national policy statements 268–72
    sustainability 79, 95, 242–3
    transport 102–8
    See also law and legislation
Green Party of Aotearoa 260, 280–1, 282
Green Star NZ 68, 74, 75, 256
'Green Tick' system 74
GreenBuild 68, 73, 74
GreenFleet 254
GreenGlobe21 256
'greenhouse effect' 9
greenhouse gases 9, 261–2
    atmospheric levels 10–11
    local authority control of emissions 260, 261, 265–6, 269–71, 272–3
    New Zealand emissions 11–12, 209, 260–2, 266, 295
    New Zealand emissions, costs of reductions 295
    See also carbon dioxide; methane; Resource Management Act
Greenpeace Aotearoa New Zealand 241, 265, 270, 280, 284, 285
group-based thinking 18–19, 24
Gulia, Milena 284

Halcrow Thomas Report 115
HALO tele-conferencing system 147, 148
halogen downlights, recessed 57–8
Halogenated Flame Retardants (HFRs) 54
'Hamilton energy blitz' campaign 154
Happy Valley 283
heating 44–5, 48–9, 85
    passive 45, 62–3
    See also water heating
Heron Motors, Rotorua 127
HERS (Home Energy Rating Scheme) 68, 74
High Court 265–6
holidays 81–2
home automation technology 61–3, 66–7, 146, 152
home renovations and maintenance 39–77, 83–4, 293
    changing how we think 66–7
    flexible spaces 58–9, 60, 76
    groups responsible for change 69–76
    'modernising' 59–60
    paradigm of 76
    planning 43–4, 58–9, 62
    resources, tools and regulations 67–9
    vision for 2020 40
household identity projects 24
houses
    carbon footprint 40, 41–2, 58, 59, 64, 71, 79, 95
    Chencha people, Africa 190
    environmental impact of living in 79–96
    home of the future 76, 81
    living in for long time 60–1, 83–4, 293
    location, and transport 60–1, 92, 105–6, 109
    moisture control 85–6
    size of 81–2
    use of electricity 12–13, 50–1, 63
    valued features 70–1
    zero-energy users 95
HOV (High Occupancy Vehicle) priority lanes 125–6
Hubbert, M. King 134
human rights 208, 209, 213
Huntly Power Station 266

Huxley, Aldous 202–3
hydroelectricity 12
hydrogen 127, 150, 152

ice ages 10–11
ice, polar 10, 187
identity projects 21
  collective 23–6
  personal 21–3
image analysis 155
imported products 132, 133, 134–5, 211
An Inconvenient Truth 19, 124, 200–1, 213–14, 231, 238–9, 286
India 127, 192, 261
  Bhopal disaster 220
Industrial Revolution 10, 190, 192, 193
'Industries of the Future' 230
information technology 144–63
  vision for 2020 145
  See also computers
Inland Revenue Department 126
in-sink waste disposal units 53
Insulating Glass Units (IGUs) 48
insulation 44, 46–7, 58, 79, 81, 85
  regulations 69
insurance, pay-as-you-drive 104
Integrated Transport Assessment 119, 120
Interface 230–1
Intergovernmental Panel on Climate Change 10, 137
International Council for Local Environmental Initiatives (ICLEI) 210
International Federation of Organic Agriculture Movements 180
International Monetary Fund 193
Internet 19, 73, 104–5, 146, 154
  interconnecting carbon-neutral networks 283–6
  online shopping 135, 138, 156–7
Internet-2 149
investment, responsible 212, 217–34, 293
  advice on 223–4
  financial performance 222–3
  investing internationally 229–31
  investing locally 227–9
  vision for 2020 217–18, 231–4
investors, four types 224–7
ISO 14001 254–5
Italy 153, 288

Japan 64, 126, 166
joinery
  aluminium vs timber 48–9
  re-using 48–9
Journal of Organic Systems 180

Kanter, Moss 234
Karangahape Road Reclaim the Streets action 283, 284

KAREN network 149
Kettle, Betty 177–8
kitchen consumption and waste 52–3
kitchens 43, 49, 51, 57, 59, 60, 66, 71, 84
  cabinetry 55–6
  consumption and waste 52–3
KiwiSaver 217, 218, 227, 232
Koanga Gardens 175
Korea 166, 170
kura. *See* schools; Te Kura Kaupapa Maori o Te Rawhitiroa
Kushner, Tony 18–19
Kyoto Protocol 14, 15, 100, 114, 116, 259, 260, 261, 281

labelling systems 50, 73–4, 255–6
Labour Party/Government 242, 277, 278, 281, 288
'land ethic' 202
Land Transport New Zealand 160
Landcare Research Manaaki Whenua 80, 253–4, 255
landfills 7, 25, 55, 88, 166, 191–2, 195, 196, 267
laundries 53–4
  rubbish disposal 53
laundry, drying 53–4, 85
law and legislation 69, 259–75
  Members' bills 287
  vision for 2020 259
  See also specific statutes and regulations
Le Corbusier, *Vers un Architecture* 87
leaky homes 84
LED lights 57, 140
Leopold, Aldo 202
Level website 68, 73
life-cycle planning 141–2
lighting 57–8, 62, 72, 84–5, 154
  recessed 46, 57–8
Live Earth concerts 286
local authorities 36, 73–7, 175–6, 267, 285–6
  Auckland, and transport 118–22
  control of greenhouse gas emissions 260, 261, 265–6, 269–71, 272–3
Local Government Act 2002 119
location of homes and jobs, and transport 60–1, 92, 105–6, 109
London 126
Long Term Council Community Plans (LTCCP) 119–20

Mahi Whenua, Unitec Hortecology Sancuary 181–2
malls. *See* shopping malls
manaaki tangata 37
Manapouri campaign 279, 288
Manchester 115
Mandela, Nelson 224

manufacturing 13, 72, 132, 133, 134–5, 155, 193, 194, 196–7. *See also* production, local; products
Manukau City Council 177
Maori 33, 82, 217, 279
Marine Stewardship Council 256
Marsden B power station 265–6
mass production 196–7
Massey University, Nanomaterials Research Centre 64
Matauranga Taiao programme 36
McDiarmid, Bunny 280
meat 90, 91–2
Medium Density Fibreboard (MDF) 55–6
melamine 55, 56
Melbourne Sustainable Living Festival 286
methane 9, 11, 12, 55, 115, 166
Metrowater 178
Mexico 194
Mighty River Power 265–6
Ministry for the Environment 14, 42, 68, 100, 255, 263, 266, 267, 268, 269, 270–1
Sustainable Management Fund 68
Ministry of Economic Development 107
Ministry of Education 36, 43
Ministry of Social Development 126
Ministry of Transport 107, 125
Mobil-gas Economy Run 94
mobile phones 146, 161, 191
Monbiot, George, *Heat* 11
monocropping 171
morality. *See* ethics
motor vehicles, internalising environmental costs 98, 102–8. *See also* cars
motorcycles, carbon dioxide emissions 117, 118
mould 85
Moxie Design Group 238

nanotechnology 64, 140, 152
National BlueGreens 282
National Education for Sustainability (NEfS) programme 36
national identities 19
National Party/Government 281–2, 288
Natural Capitalism 244
natural resources 135. *See also* specific resources
Natural Step (TNS) Framework 236, 244–5
Phoenix Organics 245–6
relating the system to business 246–52
tools to support delivery 253–6
Netherlands 122
Network for Climate Action 284
New Zealand Building Code 69
New Zealand Business Council for Sustainable Development 243
New Zealand Business Ethics Awards 231, 232

New Zealand Centre for Business Ethics and Sustainable Development 231
New Zealand Coastal Policy Statement 268, 269
New Zealand Green Building Council 68, 243
New Zealand Management 231
New Zealand Superannuation Fund 217, 225, 232
New Zealand's Climate Change Solutions 260
New Zealand's Greenhouse Gas Inventory 1990–2004 115
news, online 155
non-governmental organisations (NGOs) 240–1
North Island 14
North Shore City Council 178
Northland Regional Council 265

OECD, *Environmental Performance Review of New Zealand* 239–40
'off gassing' 55
oil 10, 196
dependence of food industry on 90
peak oil 134
rising price of 132, 134, 135, 137
oil-based products 132, 133
operations research 152–3
Oram, Rod 256
organic agriculture 164–83
certification systems 255–6
vision for 2020 165
organic gardening 169–70, 174–9, 293
Organics Aotearoa/New Zealand 180
organisational actions 293–4
identity projects 24–5
Otahuhu C power generation plant 272
overconsumption 79

Pacific Centre of Sustainable Communities 181
packaging 13, 52, 81, 87–9, 141, 158, 192
paint residues 49
Pakuranga, Permaculture demonstration home 177–8
paperless office and hime 155–6
parking 120, 121
conversion of parking spaces 120–1, 140
parking levies 103–4
Parliamentary Commissioner for the Environment 210
Parmalat 221
particleboard 55
passive cooling 46, 62–3
passive heating 45, 62–3
peak oil 134
Perfluorocarbons (PFCs) 54
Phoenix Organics 245–6
photography, digital 156

photovoltaic panels 63, 64, 65–6, 72, 140
plastic bags 22, 88
plastic tables 196–7
plumbers 72
polar bears 10
policy. *See* government policy
political activism 276–91, 294
 carbon-neutral networks 282–3
 democratic rights, maximising 287–8
 history 27–80
 interconnecting networks 283–6
 vision for 2020 277
 voting 280–2
pollution 100, 102, 103, 122, 133, 134–5, 186, 192, 193, 236, 237, 254–5, 262–3, 267. *See also* 'contaminant'; stormwater management; and specific pollutants
polystyrene insulation 47
Polyvinyl Chloride (PVC) 54, 67
positive feedback loops 11
poverty 19, 186, 193, 194, 203, 208–9, 217
power stations. *See* electricity generation
production, local 79, 81, 89
products
 climate neutral 230–1
 development cycle 155, 255
 durability 191–2
 hand-made 195–6
 imported 132, 133, 134–5, 211
 life stages 42
 locally-produced 132, 186, 214, 240
 low carbon-emitting 157–8
 mass-produced 196–7
 unsustainably produced, boycotting, 211–12, 221
 See also manufacturing
Promina Group 227
property development, and transport options 120–1
property professionals, and creation of carbon-neutral buildings 71–3
protesting. *See* political activism
public service, carbon-neutral 242, 278, 281
public transport 81, 94–5, 102, 105, 106–8, 109, 260
 and shopping malls 135, 138
 Auckland 114–16, 120, 121, 123–5, 128, 129
Pukekohe 168, 171

QE2 Trust 254
Queen Street, Auckland 122–3

Radical Routes 285
Radio Frequency Identification (RFID) 158
rail transport 105, 106–8, 135
 Auckland 114, 115–16, 123–5
 carbon dioxide emissions 117, 118
 electric 124–5
 light rail 124, 125

rainforest. *See* forest, destruction
rainwater collection 29, 34, 52, 72, 86, 178
real estate agents 66–7, 72–3, 74
Reclaim the Streets (UK) 283
recycling 52, 53, 72, 74–5, 79, 87, 88, 89, 141
regional councils 263, 264. *See also* Northland Regional Council; Taranaki Regional Council
'Regional Food Economy' programmes 180
regulations 69, 267–8, 273
religion 190, 202
renewable energy 12, 86, 124, 241, 242, 260
 and public transport 124–5, 128
 shopping malls (future scenario) 140
 See also biofuels; passive cooling; passive heating; solar energy; wind power
renovation. *See* home renovation
Resource Management Act 1991 118, 135, 259, 260, 262–4, 266, 267, 268, 271, 272, 273
Resource Management (Climate Protection) Bill 2006 259, 260, 267, 269, 271, 272–3
Resource Management (Energy and Climate Change) Amendment Act 2004 118, 259, 260, 264–6
resource sharing 82–3
'respect', in context of *Earth Charter* 207
Responsibility Scorecard 227–8, 233
Responsible Investment Association 218, 222, 223, 232. *See also* investment, responsible
reuse 48, 54, 56, 72, 74, 76, 83–4, 141, 158, 198
road pricing 103–4, 107
roads 98, 99, 100, 101, 102, 103, 105, 106, 107
 smart 160–1
Robèrt, Karl-Henrik 244–5
Robinson, Sir Dove-Meyer, *Robbie's Rapid Railway Plan* 116
Rodger Spiller & Associates 227
Rome 126
Royal New Zealand Forest and Bird Protection Society 279
rubbish disposal. *See* waste disposal

Save Happy Valley Coalition 283
SchNEWS 285
schools 23, 28–38, 176, 293
 transport to 92–3, 94
 vision for 2020 29–30
sea level changes 10–11
 New Zealand 14, 17
sewage treatment and disposal 139–40, 166, 170, 176
shipping, hydrogen-powered 150
shopping 25, 61, 62, 87, 89, 110, 131, 133, 135
 online 135, 138, 156–7
shopping malls 131–43
 vision for 2020 132
Siberian permafrost, melting 11

INDEX

smart meters 153–4
smart roads 160–1
SmarterHomes 67–8, 70, 73
Smith, Nick 281–2
social networks 21
social transformation 205. *See also* Earth Charter
Socially Responsible Investing 225
soil 165, 166, 167–70, 171, 172, 176
solar energy 29, 45, 79, 81, 95, 152, 230
  water heating 45, 51, 63, 69, 72, 86
Solid Energy 283
'solution seekers' market 238–9
South Island 14
Soviet Union 174
species loss 10, 237
spending, ethical 211–12, 292
spirituality 22, 194, 200, 202, 205
standards 254–5, 256, 263, 264, 267–8, 271
status quo, and habit 20–21, 25, 66
Stern Review of the Economics of Climate Change 101, 187, 241–2
storms 10, 14. *See also* cyclones
stormwater management 108, 133, 139, 178
'Stratford Inquiry' 269–71
Strong, Maurice 205
sun orientation 45
supermarkets 13, 52, 88
sustainability
  and *Earth Charter* 208–9, 211
  as state of mind 20–3, 197–8
  government policy 79, 95, 242–3
sustainable buildings 42, 43
  groups responsible for change 69–76
  resources, tools and regulations 67–9
  See also carbon footprint – of houses; home renovation
Sustainable Business Network 243, 252, 253
sustainable development, definition 262–3
Sustainable Electricity Association NZ (SEANZ) 65
'sustainable management', in Resource Management Act 262–3, 264–5
sustainable/responsible business 221–3, 227–8, 235–57
  tipping point for business 237–43
  vision for 2020 236–7
Sustainable Responsible Investment 223, 232. *See also* investment, responsible
sustainable schools. *See* Enviroschools Programme
Sustainable Wine Growing 256
Sweden 124, 244
Swedish Institute for Food and Biotechnology 91
'Swift' microturbines 64
Switzerland 288

tapware 51–2, 72

Taranaki Regional Council 270
taxation. *See* carbon tax; 'fart tax'; fuel tax
Te Kura Kaupapa Maori o Te Rawhitiroa 35–68
teabags 87
technology, home automation 61–3, 66–7
teleconferencing 147, 148–9, 293
tele-entertainment 149
tele-tourism 149–50
tele-working 62, 147–9, 150
temperatures. *See* global warming
three-dimensional Building Information Modelling tools 43–4
timber
  cabinetry 55–6
  joinery 48–9, 54
  non-toxic construction material 54
  treated 75–6
tipping point 186–7, 236–43
TNS Framework. *See* Natural Step (TNS) Framework
toilet cisterns 52
tomato sauce 91
Toolkit for Climate Action 284, 285
tourism 238, 239, 242, 256
  virtual 149–50
Toward a Sustainable World: the Earth Charter in Action 210
toxic building materials 54–6, 72, 75–6
Toyota Prius hybrid 94–5
TradeMe 87, 157
traffic monitoring 159
training providers 73
trains. *See* rail transport
Transit New Zealand 159
trans-national carbon neutrality network 284
transport 92–5, 98–112, 194, 292
  Auckland, reducing carbon burden 113–30
  carbon dioxide emissions 13, 92, 93, 94, 98, 99–101, 114, 115, 116–8, 253–4
  carbon-related management systems 253–4
  government policies to reduce carbon emissions 102–8
  impact of rising oil prices 135
  individual measures to reduce carbon emissions 109–10
  intelligent systems 159–61
  vision for 2020 98
  See also cars; cycling; public transport; shipping; traffic monitoring; travel; walking
travel 149
  air 13, 80, 81, 95, 99, 147, 148–9, 150, 238
  household 60–1, 81, 92–5, 99
travel miles 240
travel plans 121–2
tree-planting 16, 165, 173, 176, 260
TUSC (Tools for Urban Sustainability: Code of Practice) 68, 75
TV3 programme, *Wa$ted* 286

309

under-floor heating 45
under-floor insulation 47
Union Carbide, Bhopal disaster 220
Unitec Hortecology Sancuary, Mahi Whenua 181–2
United Kingdom 12, 95, 115, 121–2, 153, 238, 240, 284, 285
United Nations 285
United Nations Charter 202
United Nations Commission on Environment and Development 262
United Nations Conference on Environment and Development 204
United Nations Decade of Education for Sustainable Development 210
United Nations Earth Summit 206
United Nations Educational, Scientific and Cultural Organization (UNESCO) 209–10, 212
United Nations Framework Convention on Climate Change 261
United Nations Millennium Ecosystem Assessment Report 204, 237–8
United Nations Principles for Responsible Investment 225
United States 11, 19, 64, 115, 123, 147, 174, 193, 204, 209, 219, 220, 221, 224–5, 238, 240, 261, 280
Universal Declaration on Human Rights 209
University of Auckland, Planning Programme 201, 213–14
urban areas
  food growing 92, 165, 174–9
  greenhouse gas emissions 12–13
  impact of reducing number of cars 108, 114
URS 68

values 21, 23
  as motivators for change 205
  competing 20–1
Values Party 280
Vanuatu 189
Vector 64
vegetables 90, 91
vehicle infrastructure integration 160
vehicle platooning 160–1
Venice 108
ventilation 44, 46, 85
visualisation, scientific 154–5
Vitra Design Museum, France 194–5
Volatile Organic Compounds (VOCs) 54, 55, 56, 67

Waitakere City Council 68, 178
walking 13, 62, 81, 92–3, 102, 109, 292
  and shopping malls 138
  Auckland 114, 115, 122–3, 128, 129
walking school bus 93
walrus 10

waste disposal
  building sites 69, 72, 74–5, 141
  domestic, sorting at source 52–4, 88
  See also composting; landfills; recycling; sewage disposal
waste generation 81, 132, 133
waste minimisation 69
Waste Minimisation (Solids) Bill 242–3
waste reduction 260
water conservation 43, 50–2, 72, 86, 178
  shopping malls (future scenario) 139
Water Efficiency Labelling and Standards (ELS) schemes 50, 51–2
water heating 51, 86
  hot water cylinder position 51
  solar 45, 51, 63, 69, 72, 86
  wetback 45
water vapour 9, 85
Watkin, Neville 127
weather. *See* cyclones; drought; floods; storms; wind
weather prediction 151–2. *See also* climate change – models
weeds 171, 172, 175
Wellington 124, 285
Wellman, Barry 284
Whiting J 272
wind 14
wind power 63, 64, 81, 140, 230
windows 46
  aluminium 48–9
wine 256
Withers, Norm 288
wood 81
wood-burning firebox 45
wool insulation 47
Work and Income New Zealand (WINZ) 229
working from home 62, 147–9
workplace identity projects 24
World Bank 193
World Charter for Nature 203
World Commission on Environment and Development 205
World Conservation Strategy 203
World Conservation Union 202, 203–4, 205, 209
World Trade Organisation 193
WorldCom 221
worm farms 169, 177, 178